EXPLORING
THE PHYSICS OF THE
UNKNOWN UNIVERSE
An Adventurer's Guide

An Investigation of the fundamental particles and their relationship to the structure of the universe.

Written especially for those curious persons who cannot wait for someone else to find the answers.

BY MILO WOLFF
Book Design and Illustrations by
Jennifer Snow Wolff

Technotran Press
Manhattan Beach
California 90266

Copyright ©1990 Milo Wolff
Illustrations copyright ©1990 Jennifer Snow Wolff
All rights reserved.
No part of this publication may be reproduced or transmitted in any form or by any means, electronic, mechanical, photocopy, recording, or any storage or retrieval system, without permission in writing from the authors.

Published by:
Technotran Press
1600 Nelson, Manhattan Beach, California, 90266
ISBN 0-9627787-0-2

Second Edition, 1994

Printed in the United States of America.

Editing by Jennifer Snow Wolff and Lan Ling Wolff
Production Supervision and Computer Graphic Design by Jennifer Snow Wolff
Production by Associated Graphic Services, Ltd., Albany, New York 12210

Library of Congress Catalog Number 88-51875

Subject Indices: 1. Cosmology, 2. Fundamental Particles, 3. Relativity, 4. Quantum Mechanics.

Exploring the Physics of the Unknown Universe
An Adventurer's Guide

Dedicated to my wife, Lie, and my sons and daughters who with assistance, patience, kindness, and sufferance during many readings, enabled me to finish this book.

Special appreciation to my son, Eric, for his careful reading of this manuscript and his many useful suggestions. Thanks to Bruce Billings, Henry Birnbaum and Audouin Dollfus for their reading and comments on this manuscript. I am also especially indebted to my colleague, Dr. Tom Gehrels and my lifelong philosopher friend David Fisher for their invaluable help and encouragement.

–Milo Wolff
Manhattan Beach, California, 1990

Exploring the Physics of the Unknown Universe
An Adventurer's Guide

TABLE OF CONTENTS

Introduction by Audouin Dollfus

Author's Autobiography

Preface

PART I: Science as We Know It Today

Chapter 1	The Scientific Tradition:	3
Chapter 2	The Tree of Scientific Knowledge	18
Chapter 3	Choosing the Sources of Knowledge	35
Chapter 4	Fundamentals of the Unknown	45
Chapter 5	On the Importance of Living in Three Dimensions	67
Chapter 6	Evaluating New Ideas	83
Chapter 7	Space, The New Frontier	93
Chapter 8	All About Waves	105
Chapter 9	Quantum Mechanics	119
Chapter 10	Particles and Electricity	145

PART II: The Future Land of the Explorers

Chapter 11	The Universe	161
Chapter 12	The Space Resonance Theory	177
Chapter 13	Applications of the Space Resonance Theory	201
Chapter 14	Conclusions and Future Adventures	229

Mathematical Appendix

References

Index

Publications by Milo Wolff

Introduction

Of all the scientists in this world, only some scientists emerge from the crowd through unusual achievement in their field of discipline to become known as experts. These men evolve freely in the world of their particular scientific discipline. All these disciplines are the blocks which create the great building of Science.

Less frequent are those prominent scientists who are able to evolve within the whole building of Science. Covering a large variety of its aspects, they can assess the building of Science as a whole, deeply from its interior.

Exceptional are scientists who can put themselves outside the building, in order to look and think about the whole achievement, completely free of any involvement or commitment with the interior. Such is the viewpoint of Milo Wolff, the author. From this perspective, he is able to raise serious questions of fundamental concern for us still remaining within.

We, scientists working within the building for edification of knowledge, are often somewhat blind. We become committed to go along with the pressure of our environment. Milo Wolff's book helps to open our eyes. Reading it, we look at science as it is, with its beauties and limitations.

The gist of this book is beautifully summarized by extracting some quotations, taken from the chapters:

"This book quickly introduces the reader to fundamental problems of physics... and the present state of ignorance as well.

"Science, the world and the Universe, as we know it, is really an enormous amount of derived knowledge, based upon a very few fundamental facts and laws, about eight... These consist of three stable particles, the proton, the neutron and the electron, plus six sets of rules for calculating: electricity, quantum mechanics, relativity and the laws of conservation of energy and momentum...

"The rules of electricity, conservation of energy and conservation of momentum have been with us for two centuries. During this time, some scientists have forgotten that we have no idea of where they come and no concept of what causes them...Few scientists are searching for answers to these questions.

"Quantum mechanics and relativity have been known a scant sixty years, and the fact we know nothing of their origin still persists... yet the number of scientists seeking to understand them has waned.

"Thirty years ago, I began to research the history of these eight problems... Now I am relating my adventure to you, and hope that you will find as much excitement as I have investigating these ideas."

A fascinating set of deep and long thoughts about basic problems in science.... A blow for intellectual freedom.

— **Audouin Dollfus**
   Head Astronomer, Observatoire de Paris, France
   President-emeritus, French Astronomical Society

Intriguing Ideas

How often does a book chip away at the frontiers of science, and yet remains readable and entertaining? This book peers into the foundations of science, logically examines the sub-microscopic particles, and relates them to the puzzles of the Universe. You are also told what science does not know, while the book is written so as to give you a feeling of participation in finding the answers. Milo Wolff is a scholar of long standing, and his ideas are intriguing and challenging to both the novice and professional scientist."

— **Dr. Tom Gehrels**
   Lunar and Planetary Lab, University of Arizona
   Editor, Planetary Science Series

A Thumbnail Autobiography

Science is an infinite collection of gadgets.

People often demand to know, "Why did you decide to become a scientist?" Well, I didn't! Instead I have always been a "fix-it man." Such people like to repair machines and gadgets so that they purr better than ever before and if they don't run, to fix them. Machines and apparatus are nice, reliable things that work hard for little cost, so I feel good when I make them run and feel bad if I have to throw them away. Some machines frequently break down and need a lot of fixing. It isn't their fault, they just weren't built right, but if you understand them, they can be put back in good health.

When I was five years old, a neighbor gave me a box of broken electrical stuff — a toy train transformer, christmas lights, doorbells, and switches. I fixed them all and installed them in my bedroom so I could operate them in my bed. Later, I used an electro-magnet to remotely control the door-lock. Later, when I was about eight years old, I went into business fixing broken, lamps, toasters, and irons for neighbors. I charged 5 cents each.

I have been reading and learning about these things all my life. The central question is: "How does it work?" This is not an easy question. In fact, you never really answer all the questions, because nearly every time you find an answer, you also turn up a new question. This work demands patience, like scientific research. This was not my intention; I only wanted to know how things worked, so I could be sure I could fix them. I guess it was my kind of security in an uncertain world where you never know what will happen tomorrow.

During the war with Hitler, The U.S. Navy needed to maintain their new secret radar that was being installed on ships to protect against submarines. They discovered I was a fix-it man and sent me to radar and loran schools for a year. After this I served for another year and a half on the flagship of the sixth fleet in China. Our ship in the harbor at Shanghai communicated with other ships using teletype cables strung between them. The poor Shanghai fisherman soon discovered the value of the copper in the cables, and began to cut them away in the dark of night. I learned about fixing cables working in a tiny boat, but I also learned about the difference in the living standards between China and the USA.

After the war, I studied chemistry, biology, and economics, but I turned out to be an electrical engineer instead! I got a job writing books on the maintenance of army radar, and also had a night job teaching radio repair at a community college.

At this time of my life I felt I should understand how everything technological worked. Lasers, computers, and other post-war technology had not yet been invented and solid state physics had hardly any applications yet, so the idea was not too unreasonable. Still, I saw my knowledge was way behind and I decided to go to graduate school at the University of Pennsylvania and study physics.

After receiving my doctorate, I began a career teaching physics and engineering abroad: Indonesia, Sri Lanka, and Singapore, interspersed with R & D at MIT and Aerospace in California. The plight of the culture-rich, technology-poor people in Asia really disturbed me and I studied hard to find ways to fix that. Eventually I realized I couldn't because the problems are social and political rather than technical.

In the 1960's, it became obvious that new technology was being invented faster than I, or anyone, or even teams of people could learn it all. I had to face the fact that I would never be able to fix everything. The dream bubble had burst.

Still, the joys of real scientific discovery are ecstatic, although infrequent. I will not forget one late night, all alone, watching a computer draw my theoretical plots of the polarization of light from asteroids. I had worked doggedly for several years on this theory, but because of a publication deadline, I was now in a hurry. Instead of carefully testing each program part separately for bugs, I had put them all together at once and ordered plots of several kinds. I didn't even expect the program to run, but to my astonishment, there were no bugs and the theory was right. I saw plot after plot duplicating known measurements by the astronomers, many of which I had not anticipated!

Sometimes everything goes wrong! At MIT I built an apparatus to measure the green nightglow of the sky, a radiation which was useful for navigation of the Apollo spacecraft because viewed from outer space, it is a glowing ring around the Earth. To make measurements I traveled to the desert of eastern Colorado where the weathermen promised me 99% dry air and clear skies. Instead that month, records were made for rain and floods. The skies were obscured. The roads were closed, and my telephone data link went down! I dejectedly went home with less than five nights of poor data.

Now I think that the best way to learn about everything is to understand only the basic laws of the universe which underlie all science and are not yet understood by anyone. There are not many of these – about six – so this is an attainable goal of great value. In this book, *Exploring the Physics of the Unknown Universe,* I have described my work and best guess of the origin of the basic laws. Are my conclusions correct? I don't know yet and it will require more study plus experiments to prove them right or wrong. I can hardly wait to find out!

There is still a nagging problem to worry about. Suppose the Universe were to break down and quit working? I don't think I could fix it.

Author's Preface

Following the Path Less Traveled.

Curiosity explains why some people, especially children, want to teach themselves about the world by experimenting with it. They begin by poking toys, and later in life, motivated by a lifelong curiosity, may develop sophisticated laboratories or learn techniques of scientific analysis.

This book is written for the science-oriented person who has has always been curious about the world around him: the person who finds joy in logical thinking and its rewards. The person who wants to know what is out there beyond the clouds, the sky, and the stars. The person who also seeks to understand microscopic things such as the behavior of light, electricity, and magnetism. The person who is unsatisfied with the complacent explanations of others.

Anyone who is dissatisfied with the current explanations of light and electricity will find upon careful examination that his feelings are right! In fact, we don't know what light and electricity are. Yes, science knows how to describe and calculate their behavior, but the mystery of what electricity and light are and how they came to be is still very much unknown.

It is unfortunate that most books and schools usually suggest that the existing explanations of science are final and that a proper student should go away satisfied, spreading his new wisdom just as it was taught. But in reality, the frontiers of science and its foundations have always been clouded with mystery and confusion. This book enables the curious and logical person to unravel some of these mysteries to their own satisfaction.

This book is not a primer of science. It is assumed that the reader has already read about many branches of science, especially astronomy, chemistry, and physics, not because they were forced upon him as a graduation requirement, but because he wanted to know those things. The material in most of these chapters is not complex or technically involved. Instead, it deals with simple matters of science which are unknown, and seeks to comprehend them. High school courses in algebra, physics, and astronomy provide the needed background.

The curious reader and I regard science as a kind of joy. It makes us feel good when we comprehend a new concept. Our highest goal is that we may once or twice in our lives feel the ecstasy when we grasp something which had never been known before.

There is a method and a purpose in this book, which reflect the way I think about life. I am still one of those curious children mentioned above. Most of my life, I have been just as curious about things we don't understand as those which science has learned. Not many authors do this; they get joy from explaining what they know, not what they don't know. The Chinese philosopher Lao Tse, said:

"He who asks a question once appears foolish once. He who never asks a question is foolish forever."

I have concentrated on examining the simple questions of basic science which have been left at the wayside. These simple questions are really the most fascinating questions of all, for they underlie the entire structure of modern science. Most of the world's famed ancient philosophers were also intrigued by the very same questions.

I have confined the exploration to only the important basic questions. In the first half of the book, I have carefully examined the hierarchical logic of physics to separate and emphasize the fundamental laws upon which everything else is built. This pinpoints the puzzles and enigmas which need to be answered to create a foundation for present-day assumptions. These fundamental laws are surprisingly few; about half a dozen, plus another half-dozen natural numbers. In the last half of the book, I offer suggested answers to questions which I myself have proposed. So that my readers may enjoy working with the puzzles herein, I have tried to show what the problems are, how the techniques of scientific thinking can be applied, and to impart to them the fascination of exploring the entire universe in one trip!

Writing about the exploration of physics is what I know best as a teacher. The fundamental things which are simple but unknown include the meaning of energy, the structure of a particle, the mysteries of quantum mechanics and relativity, as well as the size and structure of the universe. Unraveling these puzzles should entail many delights, trials, frustrations, occasional successes, and much hard work.

The correct path to truth is usually filled with obstacles such as politics and greed. But the greatest obstacles are tradition and vanity. Famous biologists Mendel and Darwin, for example, had views acclaimed as correct, but history has recorded that Mendel might have made up some of his statistics instead of measuring them, and Darwin suggested that a giraffe got his long neck by generations of stretching. Both of the men saw a piece of the right picture but both also made mistakes. You will find that this book challenges tradition in many places, but I have done my utmost to select the challenges with the logical objective of finding correct answers.

These are my rules for exploration:

Rule 1. Know what others have done before you, and use it to go further. Recognize that most of it is probably right, but don't place complete trust in what others say; Skepticism is healthy for explorers.

Rule 2. Always leave room for error. Be willing to throw out your cherished concepts, if need be. It is a mistake to fall in love with a concept. Be willing to challenge your own traditions as well as those of others.

Rule 3. Accept both success and failure as part of the natural process of exploration.

In the last chapters I present my own answers to many of the fundamental questions of science which are raised in this book. These answers are based upon a model of the fundamental particles – a model whose structure is hidden within the elusive fabric of space itself.

Perhaps the reader will already have his own views before he reaches the end. He is asked to join the exploratory adventure and be the judge. I will be interested to learn the reader's ideas if (s)he wants to share them with me in a letter.

— **Milo Wolff**
1600 Nelson Avenue
Manhattan Beach, CA 90266

PART I:

Science As We Know It Today, Chapters 1-10

These chapters describe the fundamental laws of Nature as believed to be true by most scientists. The emphasis is to carefully understand the foundations of knowledge so that the reader can explore further and determine their causes and origins. I have made a special effort to identify only those laws which underlie all the rest of science. These laws are surprisingly few, about half a dozen, and together with another half-dozen numerical properties of the Universe, they can be used to obtain every other scientific rule known to science.

Knowledge beyond the fundamental laws does not exist. This is the curtain through which you the pioneer adventurer wish to see. I try to help you by describing some of the techniques of thinking and investigation which have worked well for famous scientists in the past. For example, Albert Einstein stated, "Imagination is more important than knowledge."

In Part II, I will describe some present efforts to look into the future, which at best is a very cloudy crystal ball.

*"Nothing puzzles me more than time and space;
and yet nothing troubles me less
as I never think about them."*

— Charles Lamb

CHAPTER 1

The Scientific Tradition: From the Occult to Reason and Back Again

The Lure of Scientific Sleuthing
The Revolutionary Beginnings of Science
Getting Ready for the Trip to the Frontier
A Short History of the Exploratory Environment
Science and Religion, B.C.-1700
The Growth of Large Scale Industry, 1700-1900
New Puzzles in Science, 1900-1940
Science Becomes Schizophrenic, 1940-Present
Telling You What Science Doesn't Know
There Are Really Only A Few Fundamental Questions
The Ideas of the Author
Mathematical Appendix

CHAPTER 1
The Scientific Tradition:
From the Occult to Reason and Back Again

The Lure of Scientific Sleuthing

What would it be like if you could possess the vision to see far into the depths of space and unravel the mysteries of infinity, to discover the truth about quasars, black holes, and neutron stars? What if you could focus on the micro-structure of the atom and understand the mechanism of the forces which hold matter together, fuel the immense energy of the Sun, make possible powerful computers, and create our fascinatingly complicated minds and bodies? These are fantasies, but it is, nevertheless, possible for anyone, using ordinary logic and study, to crack some of the puzzles of astronomy and microphysics.

This book is a guide for those sleuths of Nature who might wish to study, understand, and possibly unravel some of the basic problems of science today. It might not be as difficult as is often suggested. This detective game has always had a guiding principle: "The rules of nature are simple." So far, this has proven to be true, and there is no firm evidence that the future will be different. But at each stage in history, the keepers of the temples of knowledge have claimed that science has grown very complicated, so that only they were capable of interpreting the rules for the benefit of laymen. But when major new ideas appear, they are often so simple that dozens of persons exclaim, "Now, why didn't I think of that?"

For centuries the landed gentry, who alone could afford the role of natural philosopher, have enjoyed investigating and pondering about the structure of the universe, the cause of gravity and other forces, the meaning of time and infinity, and other questions associated with the enormous dimensions of space. In more recent decades, they have probed questions concerning the nature of micro-particles such as the electron, proton, neutron, and other exotics, and the phenomena which they cause such as: the quantum theory, electricity and magnetism, and nuclear energy. Many of history's famous scientists were men and women of title, who studied on their own. They contributed to the advance of scientific knowledge and their names remain honored. Now, the advent of the technological society and relative wealth for nearly everyone, makes it possible for you, the curious reader and amateur astronomer, to become a natural philosopher, even if you don't have a royal title and land. Why? Because you have the time to learn and think, just as they did.

Records of history show that people have always been curious to know the secrets of Nature. What lies beyond the stars? What is the smallest thing of matter? Their interest often focused on features of intense energy such as thunder, lightning, powerful winds, and the life-giving radiation from the Sun. This may have been motivated by a desire to obtain power that could conquer enemies and remove dangers. Apparently, the human animal has always recognized that knowledge is power and sought it avidly. In practical matters, the ancients learned a great deal by careful study of their environment, so that the practice of engineering in China, India, and later Europe, made early and steady progress through the centuries. Basic science, on the other hand, has only advanced rapidly in the last few centuries.

The Revolutionary Beginnings of Science

Philosophical science, based on the laws of physics, did not begin until about 1500 A.D. Three thousand years ago, some ancient Greeks thought that they had the answer to the problem of composition of matter: Everything was made of fire, water, air, and earth. Like most speculative viewpoints before the 17th century, this one wasn't backed up by experimental evidence but there were a few good guesses. The philosopher Democritis used the word *atomos*, meaning "undivided" to describe the fundamental units of matter from which we derive the word *atom*.

The effect of social ideas on scientific thinking is illustrated by a Chinese philosopher, Ao Li (c. 780-844) who wrote, "I should like to understand the meaning of the words: The accumulation of knowledge lies in the investigation (ko) of things (wan ku)." He thought this over deeply and came to the conclusion that both investigation and knowledge were desirable objectives, as follows:

> "Knowledge having been made complete, there is total sincerity in thought. Thought being sincere, the mind is then rectified. The mind being rectified, the person becomes cultivated. The person being cultivated, the family is then regulated. The family being regulated, the state is then rightly governed. The state being rightly governed, the world is then put at peace. It is in this way that one can form a trinity with Heaven and Earth." (A History of Chinese Philosophy, by Fung Yu-lan. Transl. Derk Bodde, Princeton U. Press).

Ao Li's point of view seems very logical but viewed in light of the history of China, his conclusions were not his own. Instead they were dictated by social attitudes: knowledge, cultivation of the mind, and a well-regulated family were already held in high esteem by the society. A well-governed state and peace in the land were prized social objectives, possibly because of their frequent absence. Ao Li's contribution was merely to suggest they might be achieved by arranging them as stepping stones.

Around the turn of the twentieth century, some scientists thought that they had reached the ultimate limits to knowledge. Newton's Laws (1687) could accurately predict all mechanical motion, and Maxwell's Equations (1883) described all known electrical and magnetic phenomena. Laws of heat and gases were well known. The chemical table begun by Dimitri Mendeleev (1869) was practically complete. All that seemingly remained was to polish up the details.

Then Albert Einstein, Ernest Mach, H. A. Lorentz, and others opened a new vista with the *Theory of Special Relativity* (1905). Suddenly there were many new things about which to think and rethink, concerning apparently well known concepts such as length, mass, time, and motion. Some new ideas, for example, the equality of energy and mass, $E = mc^2$, did not have their full impact for many decades to come. People thought relativity was complicated. Actually it is very logical and simple. What made it appear difficult is that the results often seemed to conflict with daily observations. Some writers tried to make it look complicated, because that provided them with an intriguing story to tell. Don't let them fool you, if you are logical there is no difficulty understanding relativity.

The second and most recent scientific revolution of this century was quantum-mechanics, which implies that matter is closely related to waves. Max Planck (1900) began it by planting a seed of enormous importance when he suggested and proved that energy is transferred in discrete amounts whose value is $E = hf$, where f is the frequency of the energy transferred and h is Plank's constant. Niels Bohr (1913) found that the constant could explain the energy structure of electrons in atoms. Louis Duc de Broglie (1925) saw the waves in Bohr's picture and calculated their size with the simple equation:

wavelength = h/momentum.

Equations defining the waves were quickly found (1926-1938) and soon most of physical science was seen to be a consequence of these two concepts: relativity and quantum mechanics.

The third revolution is yet to come. It is tantalizingly close and should answer the important questions: Where do relativity and quantum mechanics come from? How are the two concepts joined? Is there an underlying explanation of them, or must we accept them separately, and just as mathematical rules? These questions rephrase the old philosopher's quest for ultimate simplicity. To find this was Einstein's last self-imposed task, which he never finished. Would you like to try?

The three questions above have never been answered, although related phenomena have been well-examined and described. This is because it is often easy to find out *what* happens, but it is *more* difficult to understand *why* it happens. It is probable that these pioneering questions have not been answered because great imagination is required, as well as scientific training. For this reason, the intelligent layman who enjoys exercising his wits and imagination may become as much an expert after some study of the problems as the professional scientist. This book will show you how to try.

In the last half century, there has been an increasing awareness that the problems of the large-scale universe and space itself may be closely connected with the microphysics of the smallest particles. This is the prominent clue which suggests that an enormous simplification of science is possible by finding the relationship.

The following chapters will follow this possibility and present the gamut of unsolved problems of cosmology and particles. You, the reader, are invited to think up new ideas to answer these fascinating questions and to work further if you are curious. Current theories which are popular with most scientists, including some of the author, will be described, so you know exactly who the competition is and what you have to do to beat them.

Getting Ready for a Trip to the Frontier

Before setting out on a scientific journey, it is important to understand the changing attitudes of people and the scientific community towards new ideas and "truth" through the centuries. Scientific attitudes have always been greatly affected by the religious, political, and economic climate around them. Idealists would often contend the opposite: that the search for truth has been pure and logical. History supports the first contention.

Scientists are people first of all, and suffer the same fears and uncertainties, as other persons. Like most of us, bread on the table, a car in the garage, and security for the future have the first priority in their lives.

Science, viewed in historical perspective, has usually followed the lead of the dominant economic, political, or religious powers of society. The value of science, the rewards for science, scientific goals, its methods and fashions have all changed according to the demands of powerful people and institutions.

The situation is not different today. If you are going to get into the game, and you want your ship to make headway, it helps to know which way the political wind is blowing. Let us peek at history as summarized in Figure 1-1.

A Short History of the Exploratory Environment

CHAPTER 1/*The Scientific Tradition*   7

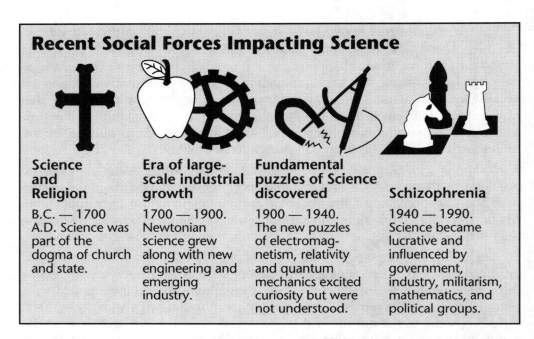

Figure 1-1. Periods of scientific history and social forces which shaped the development of scientific goals and research.

Science and Religion, B.C.-1700

Most people stood in awe of natural phenomena which they did not understand and created explanations of it to satisfy their desire to understand. Others studied it closely. Some entrepreneurs went a little further and turned it into a business. Court astrologers and palm-readers told their lords about the future. Witch doctors of the tropics, priests of the West, and medicine men of the North put their knowledge to work for their customers or patients who were able to pay the fee.

How were these people regarded within their society? It is a mistake to think that they were regarded as charlatans, even though today we might be amused at the state of their understanding using our hindsight. On the contrary, up to the present day, mystics occupy high positions in their societies. The royal courts of medieval Europe maintained astrologers and doctors of high rank. They were trusted and respected. Today's astronomy evolved from the ancient art of astrology. Some of these professions have disappeared into history, but the support of science is still similar. As the boundaries of knowledge change, the prophets, doctors, and professors adjust their realm of practice to fit the new information. Human behavior has not changed, only knowledge has changed.

The Catholic Church Agreed with Arisotle. Explanations of science at the exploratory frontier often end up as religious doctrine. In Europe, the Catholic Church subscribed to the teachings of Aristotle who saw the Earth as the center of the Solar System with the Sun, Moon, stars and planets revolving around it. He held that all heavy bodies have a "natural" tendency to move towards the center of the Universe (the Earth). All other motion was "violent" motion because it contravened the tendency of bodies to sink to their natural place, the center. The acceleration of falling bodies was explained on the grounds that they moved more "jubilantly" as they approached home — somewhat like a horse reaching his stable. The heavenly bodies were exempt from the natural tendency because of the intervention of a Supreme Intelligence or Prime Mover. Except for falling bodies, things moved only when and as long as effort was expended to keep them moving. They moved fast when the mover worked hard, and stopped when the mover stopped.

On the whole, Aristotle's theory of motion squared well enough with common experience and so his ideas prevailed for more than fifteen hundred years. Eventually men began to discover discrepancies with measurements of real motion, such as the case of the arrow which should have fallen when it left the bow but stubbornly kept moving. These paradoxes were met with ingenious modifications, in order to keep the philosophy alive.

Galileo Galilei (1564-1642) was born of noble parents who educated him in medicine, science and mathematics, and spent most of his life as a university lecturer. His ideas on the nature of motion and gravity challenged the mistaken concepts of Aristotle, and more importantly, his opposition to Aristotle, and thus the Church, changed the climate and direction of scientific thought in the whole of Europe. He was the principal figure in this revolutionary drama of changing ideas. He brought together diverse observations of the motion of bodies and firmly established the mathematics of motion, based upon the modern concept that, "a body continues its motion in a straight line until something intervenes to change it." He also made major contributions to acoustics, hydrostatics, and designed the first pendulum clock, the first thermometer, and a refracting telescope with which he became the first person to see the moons of Jupiter.

An enormously popular lecturer, his future seemed assured, but in 1632, when he attacked the Aristotlean concept of the Solar System arguing that the Sun, not the Earth, was at the center, he found himself contradicting Roman Catholic doctrine. Consequently he was tried for heresy, forced to recant, and sentenced to house imprisonment for life. A well-remembered quotation from his writing is ironic: "In questions of science, the authority of a thousand is not worth the humble reasoning of a single individual." He died the year Newton was born.

Isaac Newton was the outstanding scientist of this period. His unique vision laid the foundations of modern physics, and changed the course of human experience for all who came after him. Born on Christmas day 1642, he fully understood the power and influence of the Church and took great pains not to cross swords with it. He was an exceptionally skilled lawyer, theologian, historian, and investor, as well as the mathematician, astronomer and physicist for which he is most remembered. He died a rich man.

Newton's voluminous private papers showed that he believed strongly in God, alchemy, and magic. Early in life he abandoned belief in the Trinity, which was the roots of his *alma mater* and employer Trinity College, in favor of other sects. But he was very careful to keep these secrets locked up, for Trinitarians were the dominant religious group.

He looked upon the whole universe and all that is in it as a *cryptogram* whose secrets he could read by applying pure thought to the evidence which God has revealed in the operations of nature. These mystic clues were a sort of philosopher's "treasure hunt" to the brotherhood of scientific seekers of which he was clearly a favored member. He spent long hours alone, reading, thinking, and writing. He left more than a million words of script.

Alchemy, theology, transmutation, the elixir of life, and other magical studies received as much of his attention as science. But his papers show that he applied the same careful, accurate, analytical thinking to them as to his famous scientific work, the Principia. Apparently it was possible for the most famous and brilliant scientist of all time to involve himself in what we now know to be superstition. Did Newton really distinguish between magic, religion and science? Or were all these matters legitimate questions for the intelligentsia of his society?

The Growth of Large Scale Industry, 1700-1900

The population of the cities grew, commerce and machine-industries multiplied, ships and the profits they brought became large, and the power of the Church waned. Not only was there no dominant social power to dictate an arbitrary course for "science," there was a profitable use for discoveries. As a result, men of this period were encouraged to seek out the principles of nature without significant restraints. There was a demand for larger amounts of machine power and larger machines. Nature was obviously powerful and the researcher of nature was quickly rewarded with knowledge that turned into gold.

It soon became apparent that there was a logical way to proceed:

Step 1. Make measurements of natural phenomena.

Step 2. Propound the simplest theory which fits the measurements.

Step 3. Test the predicted measurements of the theory against more measurements.

If step 3 was positive, scientists felt that they had found "proof" of the truth of their ideas. These steps became known as the "Scientific Method." One discovery followed another, and in a short two hundred years, the basis of modern science and engineering was laid out in textbooks and practice.

New Puzzles of Science, 1900-1940

During the previous period most of the discoveries of science involved measurements in the range of human experience. That is, dimensions that we could see with our eyes (about 10^{-6} to 10^6 kilometers) and forces that we could feel with our muscles and sensors (about 10^{-6} to 10^6 kilograms) and so on. Their application to engineering was highly successful and people felt that the Scientific Method was verified; logic and reason had triumphed. Ideas of science were regarded as reasonable because they usually agreed with our senses and when they occasionally didn't we could persuade our senses to agree with science.

But then three new disturbing phenomena were discovered which could not be ignored: the laws of electricity propounded by James Clerk Maxwell (1883), the ideas of Special Relativity (Albert Einstein 1905) and Quantum Mechanics (1920-1930). These led to great confusion because experiments and conclusions frequently disagreed with our human senses. The question was asked, "How can electric and magnetic forces act across empty space?"

Some philosophers argued logically that we had no reason to expect agreement because these were all taking place at dimensions smaller or larger than the range of human experience. Quantum mechanics, and the "particles" it ruled, always involved micro-dimensions (10^{-6} kilometer and smaller). Relativity appeared when velocities approached the speed of light and distances were those of interstellar space.

Most scientists reacted by creating electric, magnetic, and other "fields" in their imagination which substituted for tangible materials to extend forces across apparently empty space. Fields and photons became the means for carrying energy from one place to another. Paradoxical behavior which could not be explained was often set aside for decades.

Confidence in the Scientific Method weakened. Quarrels arose over the meaning of quantum mechanics. Religion and the occult reappeared as explanations for quantum paradoxes. Some simply dismissed the necessity of understanding and instead claimed that mathematics was all that mattered.

Interestingly, the laws governing these three phenomena are not at all difficult to grasp. They are little different than gravity which most of us feel we understand because we can feel it. But illusions of great complexity grew out of the confusion, and priesthoods arose to interpret the enigmas.

Science Becomes Schizophrenic, 1940-present

The uncertainties led to a split of the philosophy of science into two camps. One camp was led by Albert Einstein at Princeton with David Bohm and Paul Dirac in England, who held that reasonable and logical explanations existed. Einstein felt that the mathematics of science should describe real things; i.e., models should exist for our thinking and nature has no "spooky" events that we cannot explain. They argued that a logical explanation could be found for bizarre phenomena such as the wave-particle duality of quantum particles and the "photon," and the changes of mass, length, and time associated with relativity. Their view was epitomized in a famous remark by Einstein: "God is subtle but not devious."

The Einstein camp slowly lost ground as the desired explanations failed to appear, and since their final defeat, known as the EPR experiment (Einstein, Podolsky, and Rosen), the views of this group have almost disappeared. The story of EPR is quite fascinating and will be described in Chapter Nine, because the argument has not faded into history; it is still active and alive. Briefly, it involves the question as to whether information can travel faster than the speed of light, or even whether knowledge can be had about events before they occur!

The other camp was led by Niels Bohr in Copenhagen, Denmark who increasingly became convinced that there was no hope of explaining the mysterious and puzzling effects of quantum mechanics. The human mind was unable to distinguish between reasonable and unreasonable behavior of quantum particles in the realm of microphysics which was far from human experience. Instead, one had to rely upon the consequences of mathematical analysis. This viewpoint became known as the *Copenhagen Doctrine* (CD). Present day consequences of the CD include proposals that the space we live in may contain many dimensions, rather than the familiar three we all experience. The "Big-bang" theory, and the quark particles, of which you have probably heard, have CD origins.

The recent dominance of the CD can be viewed as a fashion of science, because there is still scant definitive evidence one way or the other. The CD point of view is supported by its disciples, the mathematical physicists. They are a well-organized group and the lectern of the universities spread their concepts. Mathematics is more easily argued than the unstructured natural philosophy of Einstein.

As the fortunes of the CD group rose, the reliance on the Scientific Method by mathematical physicists waned. This created a schizophrenia in science because the experimental physicists still found that their success and careers depended on the Scientific Method, while theoreticians found themselves on a popular stage by announcing bizarre mathematical theories of the universe. Various authors suggested we live in a universe of 4 dimensions, then 5, then 11, and that multiple worlds exist simultaneously, all uncommunicating and unknown to each other. The mind-boggling idea has emerged that the universe was created by a primeval blast of energy of unimaginable size — almost infinite. This "Big-bang" occurred when the size of the universe was zero! Mathematics is the guide and crystal ball which leads to these strange conclusions. The presses print it all. This is certainly not a scientific method, but it is very popular today. It is a new religion of science.

Telling You What Science Doesn't Know

In this book, a goal is to quickly introduce the reader to the fundamental problems of physics. To this end, the present state of knowledge in fundamental areas is reviewed, not to just describe what *is* known, but to make clear what is *not* known. In short, the curious reader has to be informed of the present state of ignorance as well. Accordingly, whenever a basic question is encountered for which present day science has no answer, it will be emphasized by labeling it as follows, for example:

ENIGMA 1-1: *What is Truth when all knowledge is False?*

Sometimes information is recognized by the science community as being very important but no one has yet found a way to obtain it. For example, everyone would like to know if there are other intelligent creatures in the universe, but the difficulties of communicating with or traveling to other stars leaves us without the information. Such desired information is emphasized by using the following label:

RESEARCH GOAL 1-1: *Is There Other Life in the Universe?*

Sometimes questions arise which seem impossible to answer because there are two or more answers which appear to conflict with each other. Or, the only apparent answer conflicts with a basic rule of physics which we think cannot be wrong. One often suspects that important concepts lie hidden behind these conflicts. These questions are labeled:

PARADOX 1-1: *How Can a Blind Man See?*

The number "1-1" means it is the first paradox in the first chapter.

Frequently there are questions which are a lot of fun or reveal new ideas and applications if you think about them awhile. I label these:

QUESTION FOR THOUGHT: *How many ways can you think of to define infinity?*

There Are Really Only A Few Fundamental Questions

Science, the world, and the Universe, as we know it, are composed of an enormous amount of derived knowledge, based upon a few fundamental facts and laws — about nine. It is these nine things which we must regard as the problem questions of the universe, because no one understands their origin. These consist of three stable particles: the proton, the neutron, and the electron, plus six sets of rules for calculating: electricity, quantum mechanics, relativity, and the laws of conservation of energy and momentum.

By making careful measurements, physicists and astronomers have accurately deduced the properties of the three particles and know that all the matter of the universe is made up of them. But what are they? How are they formed? No one knows. However, by just knowing their properties and the rules of electricity and quantum mechanics, it is possible to clearly understand the formation of atoms and molecules, to create new materials, and to accomplish all the remarkable things which have led to lasers, micro-chips, fantastically powerful computers, and all the other machinery of modern technology.

There are other known "particles" besides these three, such as the lambda, kaons, and muons, and others known from high-energy experiments, but none are stable since they decay to either protons or electrons. Only the stable particles contribute to the mass of the universe.

Some, such as the quarks and photons are mathematical entities without independent existence. These particles are discussed briefly in Chapter Ten. The rules of electricity, conservation of energy, and conservation of momentum have been with us for two centuries. During this time, some scientists have forgotten that we have no idea of where they come from and no concept of what causes them. The rules are simply accepted as "truth." Few scientists are searching for answers to these questions. This is amazing, since the basic nature of these problems makes the answers potentially the most useful things we could know.

Quantum mechanics and relativity have been known a scant sixty years, and the fact that we know nothing of their origins, still persists as a nagging question in scientific minds. Yet the number of scientists seeking to understand them has waned. This is perhaps the weakness of human behavior expressed by: "Out of sight is out of mind."

The Ideas of the Author

Thirty years ago, I began to research the history of these nine problems. The first reason was because I could not understand quantum mechanics. I could see no logical reason for the simple equation of a deBroglie wave,

$$\lambda = h/\text{momentum}$$

It really bothered me that I could see no origin of this experimental rule and it continued to bother me until about 1982, when I found what I hope is the answer. Before that time, I thought I understood gravity, energy and momentum, but of course I did not. As I searched and tried to put things in a logical order, it began to appear that all of these nine problems were inter-related.

Now I am relating my adventure to you, and hope that you will find as much excitement as I have investigating these ideas. I have tried to clearly present the evidence and the problems, so that if my explanation is wrong, perhaps you will be able to find the right one.

My explanations are not presented until Chapter Twelve, where the concept of a *space resonance* is introduced. This new concept is a pair of spherical waves one of which is outgoing from a point and the other is ingoing. They move in the fabric of space with the velocity of light. Together they display properties, such as charge and mass, just like the electrons, protons and neutrons. Two *space resonances* interact with each other according to proposed *properties of space*, and lead to the rules of quantum mechanics, relativity, conservation of energy, and momentum. Here is a possible answer to these centuries-old, basic questions of natural philosophy. A lot of checking and experiments must be done to find out if they are correct.

Should they turn out to be correct, the present day assumptions and unknown constants in science will be reduced in number. The new assumptions, all properties of space, are three in number. In contrast, the formerly unknown origins, constants, puzzling phenomena, and rules of physics now number about fifty. These fifty enigmas are replaced with about three new ones. If true, the result is a new field of research to answer the question: "How can we understand the nature of space?" This prospect is very exciting, but as science always turns out, there is never a last question. Every answer reveals more questions.

Mathematical Appendix

I want to present ideas and information about physics and cosmology and to create new avenues as part of the fun of sleuthing. This goal is often hampered by the mathematics necessary to prove some concepts. Even though mathematics is often interesting, it isn't fun for everybody. So in the text, conclusions which need math will be merely stated and a proof given in the appendix, which you can examine if you wish.

"If you tell yourself something over and over again, right or wrong, it becomes intuitive."

– Sidney Coleman

CHAPTER 2

The Tree of Scientific Knowledge

Einstein's Goal Revisited
Three Realms of Measurement
The Puzzle of Length, Time and Mass
Energy Exchange, the Heart of Measurement
Fundamental Laws of Nature

CHAPTER 2
The Tree of Scientific Knowledge

Studying science is different than studying history or language. Nature has only a few basic objects and rules which determine all the great variety of things we find in the universe such as stars and sapphires, flora and fauna, pinball games and planets, and you and me. The variety occurs because of the different arrangements which are possible using a few simple building blocks. Protons, electrons, and neutrons make up 92 different atoms and these atoms make countless molecules which in turn constitute metals, crystals and organic matter. It is like brick-work: many different kinds of buildings can be built of the same bricks.

Science has filled whole libraries with details of the arrangements and classified them into disciplines such as chemistry, ceramics, glues, semiconductors, etc. This knowledge can be arranged into a tree-like structure. At the roots below, are the fundamental objects and the laws for combining them. Above, combinations of them spread out into endless branches of chemistry, biology, applied electricity, engineering, agriculture, the weather, pottery, food, housing, and so on.

We understand how atoms and molecules combine because we know how to use the laws of combining basic objects. But we generally don't understand the objects, nor how they are formed or relate to each other. And we don't understand where the laws come from. It is the simple things we don't know! These are the things we want to search for in this book.

An important goal of this book is to help you the reader join the game of scientific sleuthing. As an amateur you may be handicapped by not having a long professional education in physics and astronomy. I intend to help by providing you with some tools for evaluating science, and by describing to you all the unsolved paradoxes that I can find. These paradoxes may be the very clues you need to pierce the veil of mystery at the frontier.

It could also turn out that your amateur status will help you rather than hinder, for you do not have biases, fixed ideas, and career investments to protect which accompany education and experience. It is certainly true that professional education can burden you with myriad details of useful science, but these may not be useful to you because they have almost no connection with the problems on the frontier.

If you succeed, it will not be the first time that a fresh outlook and the imagination of youth have raced beyond the wiser and older grey heads of the experts.

Einstein's Goal Revisited

This book will not follow the Copenhagen Doctrine (CD) of total reliance upon mathematics. Instead, it will be assumed, following the ideas of Albert Einstein and Isaac Newton before him, that scientific evidence exists which our instruments and senses can find. When we find the evidence, it will allow us to understand and interpret all the rules of Nature without accepting paradoxes as inevitable. This view will be termed the "Geometric Method" (GM) because our minds tend to create physical models of things we see, hear, or feel. The CD view is to algebra as the GM view is to geometry. It is interesting that many mathematical proofs can be obtained by *either* of two methods; one algebraic and the other geometric.

There are two reasons for following the GM road. Most important is the existence of large amounts of evidence concerning microphysics and the cosmos gathered over the last century which have not been utilized by either of the two camps. Much of it has been just swept aside as interesting but useless paradox.

The other reason is that although quantum (micro-sized) behavior and relativistic (cosmos-sized) behavior are outside the range of human experience, scientists still describe and calculate them using the same human units of length, time, and energy. The objects we find and their behavior are strange, but we must still think and reason about them. I have confidence that new young thinkers have this imagination and will find the way to understand in accordance with the Geometric Method.

Three Realms of Measurement

Let's begin. You and I must agree to arrange our thoughts in the same way at first so we will both understand the same important concepts and communicate with each other accurately. This requires us to agree on certain definitions of words and ideas.

First, we must deal with the fact that our human senses have limitations. We are poorly equipped to observe things in the world around us which are very large or very small. Let us then agree upon three ranges of size of objects and quantities to describe them as shown in Figure 2-1.

The most familiar is the **human realm.** It encompasses lengths, speeds and masses which we can see with our eyes, feel with our muscles and nerves, and hear. This range has been extended with microscopes, telescopes, and other instruments. Like many definitions, there are grey areas at the edges which fade

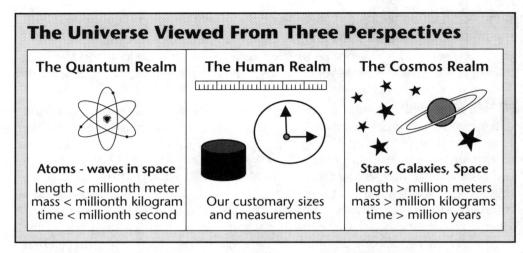

Figure 2-1. Nature has divided the Universe into three realms.
The three realms of science encompass three different ranges of size of objects and their measures.

off to the next range so we cannot say exactly where one begins and the next one ends. Roughly, let's say these ranges go from a million kilometers to 1/millionth of a kilometer for length which is about the distance of the Moon and the size of a molecule; a million to 1/millionth of a kilogram or the mass of a skyscraper compared to an ant; and a million to 1/millionth meters per second for speeds. It works out that way because meters, kilograms, and seconds *are* historical choices, which from experience, people placed near the center of our sensory range capability.

The second range is the **quantum realm** which includes everything *smaller* than the human realm. Science has found, perhaps not too surprisingly, that physics is different here. This realm is dominated by the laws of *quantum theory,* which reduced to its essence, says that particles and their motions appear to be some sort of waves. We don't know exactly what kind of waves, but we have fairly good mathematics for calculating their behavior. Particles behaving like waves seems very odd to us because we have had no experience with such things. Luckily, the rules of behavior are not complicated, but you cannot compare the waves to anything in the human realm. It is very important to remember that in the quantum realm one is always dealing with one particle at a time or just a few. It is the *single* particles which look like waves.

The third range is the **cosmos realm** which includes everything *larger* than the human realm: large distances, high speeds, and large objects. Especially important is the speed of light (3×10^8 meters per second) which is the greatest possible speed of any object. Again, scientists have found new and different behavior. Three new phenomena are involved: *Special Relativity, General Relativity,* and the *Hubble Rule.* This realm is characterized by changing mass, length, and time depending on the speed of the object with respect to the person or instrument making the measurement. Such changes are large when the relative speed is comparable to the speed of light, but are hardly noticeable in the human range. That is why they were not discovered until the early 1900's using accurate instruments.

In the cosmos realm astronomers have found a fact which is perfectly obvious and logical, except that most of us never have any reason to notice it, so when told about it we think it strange and odd.

> **FACT:** *In the cosmos realm, there is no way to choose any object, place, or point as a unique fixed coordinate system.*

There is nothing seen in the visible universe to suggest to astronomers that they can find a unique origin for a measuring system — not on the Earth nor anywhere in space. As a result it is impossible to say that anything is holding still, or that anything is moving with a particular velocity. Does this not seem strange? Especially with traffic cops around who insist you are going faster than official speed! Why is the traffic cop so illogical? He doesn't understand it, but what he really means is that you were going too fast *with respect to* his stretch of road, next to the tree he hides behind. That is the heart of the matter: all speeds must be measured *with respect to* some other object. The speeds of cars are usually measured with respect to the surface of the Earth. The Moon's speed in orbit is measured with respect to the Earth and other planets. The Earth's speed in orbit is measured with respect to the Sun and other planets. This continues *ad infinitum* because nothing can be found to be fixed. Obvious, when you think about it!

Why doesn't everybody notice this? Because we are used to walking and standing on the Earth. We imagine without thinking that this is a fixed place. Astronomers have known for two hundred years that it wasn't fixed, but paid no attention to it. This simple concept is so fantastically important it needs emphasis. *It is fantastically important!* Why? For two reasons. First, because the world's finest scientists labored for about 20 years at the turn of the century to gather evidence to prove this simple idea was true. It became the logical basis for the Theory of Special Relativity (Einstein). It is *all* you need to know to create that theory. It is the *only* building block needed. Notice that you, the amateur scientific sleuth, had only to grasp this idea by thinking logically about how to measure speeds and realize there were no mileposts nailed up in outer space or anywhere in the universe. This is a beautiful example of the power of logical thinking. Further, it is an example of how simple and subtle we have found Nature to be.

Relations between Realms. These three realms are not unrelated. The world, as we see it in the human realm, is made up of large collections of tiny objects from the quantum realm. It is just like a printed photo which may look like one image until you get up very close and discover it is made of many tiny dots. For example, a diamond stone apparently having very smooth facets is really a collection of quantum wave-centers held together in a lattice by the quantum rules for waves. Each wave-center is comprised of six neutron, six proton, and six electron wave-centers. These 18 centers are held, also by the quantum rules, in a unique three dimensional matrix, which identifies it as a carbon atom. In the quantum realm, it is not smooth at all!

Our human senses are never able to detect single atoms or even single molecules. What we imagine to be particles are always collections of quantum wave centers. If we want to be accurate and logical (necessary for the science game) it is important to realize that *we humans often think we see particles but never do*. When we measure objects in the human realm, we are actually measuring large groups of atoms or molecules or wave-centers. The observations we find in the human realm are different than quantum, but still their origin is quantum; that is, human observations are the averages of many tiny objects which follow quantum rules.

Right now is the time to mention that the rules of relativity in the cosmos realm also apply to the human realm and the quantum realm, but they are seldom noticed because human realm speeds are small by comparison. However, there is one noticeable case of great importance: magnetism. Magnetic forces are the result of the slow motion of electric charges, but because the electric forces are enormous, the ordinarily negligible relativistic effect becomes appreciable. We never experience these enormous electric forces because (+) and (-) charges are always present, nearly always balancing each other's force. Small relative motion of charges can upset this balance to produce differences which we name magnetism. The rules of relativity will be discussed later.

The Puzzle of Length, Time and Mass

In our complicated lives we measure all sorts of things, for example: bank accounts (dollars), tire sizes (inches of diameter), speed (meters per second), our weight (kilograms), paint colors (frequency of light), etc. But it has long been realized in science that all of these different units can be reduced to just three basic ones (four are needed if you include electric charge). Many choices for the basic three are possible, but the usual choices are **length**, **time** and **mass**. All the rest are combinations of these three. Since this book intends to think about basic problems of Nature, it is very important to understand the three basic units as suggested in Figure 2-2. What determines them?

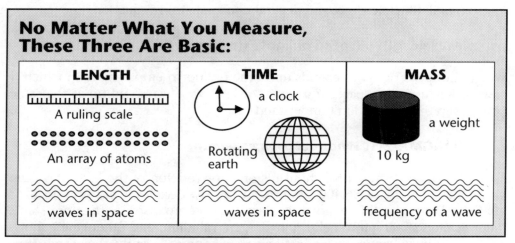

Figure 2-2. Do we understand what these really mean?
   The three basic units of measure are length, time, and mass. They are so familiar to us, we are almost certain what they mean. But when we look closer, the meanings disappear and we are not sure at all!

Think about **length.** What is it? Engineers who want accurate measurements take them from standard bars of platinum which have fine lines scribed on them. There is a standard meter-bar kept in Paris for this purpose, and copies are kept in major laboratories of the world. As an alternative, a world agreement has been made to use a certain wavelength from krypton gas when it radiates, like a "neon lite." These waves can be turned into length standards by resonating them (back and forth reflections) between mirrors whose separation is then an exact multiple of the wavelength.

These definitions of length seem quite clear and understandable, but *do* we understand? Suppose we magnify the meter bar until we are observing the platinum atoms held together in their lattice. Now we notice that nothing is very solid anymore. We can count the number of atoms between the two fine lines, but there is a great deal more empty space than atoms between them. So, what determines the standard length? Is it empty space? What is empty space? Suddenly you remember that the rules of quantum hold the atoms exactly spaced in their lattice, so you decide it must be the quantum rules which determine length. Unfortunately, the origin of the quantum rules is quite unknown — except that they refer to waves traveling in empty space. So no matter which way you try, you come to the same conclusion: length is easy to define and find, but its determination has something to do with empty space, which you don't understand.

Let's not give up. Try the other definition of length: the wavelength of the krypton radiation. Here too, you depend upon a defining equation:

wavelength = (speed of light)/(frequency)

where the speed of light depends upon the unknown empty space in which it travels. Again empty space! You sigh, "At least we always end up at the same place, even though we don't understand it."

ENIGMA 2-1: *What determines length?*

Think about **time.** Throughout history, one rotation of the Earth has been defined as one day or 86,400 seconds. In the last decades it has been realized that the Earth should be slowing down, and comparison with more stable "atomic clocks" (krypton, helium, hydrogen, rubidium and other gases) shows that Earth time changes by a second every few years. Recently, the atomic time standard has been adopted.

What determines time? We can only guess that the mechanism of a beating atomic oscillator is similar to that of an oscillating organ pipe or a violin string. The rate is proportional to the length of the pipe or string and inversely proportional to the speed of the wave which travels through them. In an atomic oscillator, the wave travels with the speed of light. We are led, once again, back to the same mechanism as the krypton oscillator which seems to depend on empty space.

Isn't there any other choice? Look at the rotation of the Earth and carefully examine how we determine exactly one revolution. We measure the beginning and end of a revolution when the position of a fixed star (not a planet) passes the cross hairs of a rigidly mounted telescope. The light which travels from the stars goes through empty space. It is almost the same result as before, except now the result seems to involve the position of the stars, as well as the empty space. Why? What could the stars have to do with time?

ENIGMA 2-2: *What determines time?*
ENIGMA 2-3: *What do the stars have to do with time?*

Finally, think about the third unit, mass or energy or frequency. Science has been making measurements of mass, energy, and the frequency of light waves for a hundred years or so. Einstein proclaimed the equivalence of energy and mass with "$E = mc^2$", and Planck said, "$E = hf$", meaning that the frequency f of a photon which transfers energy between two charges is proportional to the actual energy. Careful measurements have verified equality of these two rules without exception, so we accept the equality as a fact of nature. Either mass or frequency or energy can be used as the basic unit, since they are equivalent to each other except for the multiplier constants c^2 or h. Our task now is to attempt to understand them. For understanding, they are not equally useful since they are measured in different ways. But before we try to understand we must know about the quantities which go into each equation.

Mass m, can be measured in *two* different ways, which appear not to be connected with each other, but always give the same answer. The first method uses the force of gravity, discovered by Newton, $F = Gm_1m_2/R^2$ where G is a constant, R is the distance between two objects, say the Moon and a spacecraft, and m_1 and m_2 are the masses of the spacecraft and the Moon. In this method, one compares an unknown mass with a known (defined) mass using a balance scale so that both masses are attracted by the Earth's gravity. This kind of mass is called *gravitational mass*. We know how to calculate it but we don't know how it works, or its origin, or its connection with other forces. It is just there.

ENIGMA 2-4: *What is gravity?*

The second method of finding mass uses another rule of Newton, $F = ma$, where F is the force, say from your arm muscle, needed to speed up the mass m, say of a rock you are throwing across a river with an acceleration a. This kind of mass is called *inertial mass* because the resistance to change of speed is what we mean by the word *inertia*. Just like gravity, we don't known the origin of Newton's inertial law, although we can calculate it perfectly.

ENIGMA 2-5: *What causes inertial forces?*

It is yet another great puzzle of nature that we do not understand why these two independent ways of defining mass give the same result. The equality of the two kinds of mass have been verified for materials made of electrons, protons and neutrons. But there is no knowledge of the gravitational properties of positrons, anti-protons or other particles because gravity is too weak to measure for a single particle and we know of no way to combine them into heavier materials. Do anti-particles have gravity or anti-gravity? Scientists would like to know.

ENIGMA 2-6: *Why are inertial and gravitational mass identical?*

RESEARCH GOAL: *Find the gravity of the anti-particles.*

In the relation Energy = hf, the frequency f is that of a light wave, radio wave, x-ray, or other electromagnetic wave traveling at the speed of light to exchange energy. Such waves always accompany an exchange of energy between two objects. Indeed, it appears that the waves are the means of exchange. It is a quantum-like exchange because one can observe hf energy units, one at a time, being exchanged in sensitive photo-cells.

It turns out that measuring frequencies instead of energy is not very practical in many ordinary situations where energy is consumed, such as frying eggs, melting steel, or racing cars, so it is rarely used, and most engineers and scientists never even think of frequency as an energy unit. Instead, practical units such as kilogram-meters, kilowatts, and calories are used to measure energy exchanged. Likewise, in commerce, mass-like units such as grams, pounds, and tons are more practical for trade and engineering.

But remember, any of these energy units can be converted to any other using the right multiplying constant, so which one you use is a matter of preference. In this book we will use the energy unit of *frequency* because it is the *only one* which describes a mechanism for the energy exchange. This mechanism is that "something" travels between one object and another with the speed of light, like a light wave. One of the two objects has its frequency (mass) decreased while the other's frequency (mass) is increased. The machinery of this exchange is poorly understood even though we know how to calculate the numerical values of the frequency (mass). But poor understanding is better than none and the use of frequency provides a beginning and hope. That "something" which travels is usually termed a "photon." Giving it a name doesn't add any knowledge but it allows us to talk about it more easily.

ENIGMA 2-7: *What is a photon?*

Time, Length and Mass in the Tree of Knowledge. If you have already studied physics or engineering, you probably know that a thorough understanding of units of measure is useful because units are commonly used to check the results of work in algebra; the units on one side of every algebraic equation must always equal those on the other side. But units are important to us for a different reason; if we can understand the measured facts at the root of the tree of scientific knowledge, then that understanding can be extended throughout the whole tree. Most of science is obtained by measuring the world around us. Our understanding depends on our interpretation of the measurements. With this in mind, this discussion can be summarized:

1) When interpreting measurements and observations, care must be taken to relate them to either the **human realm**, the **quantum realm**, or the **cosmos realm**.

2) Basic units of measurement are **length**, **time**, and **frequency/mass**. Other units can be obtained from these three. We do not understand the origin of these, but examination shows a possible connection with empty space.

3) **Mass, frequency** (of a photon), and **energy** are equivalent units except for multiplying constants. **E = mc^2 = hf.**

Energy Exchange, the Heart of Measurement

Nature appears to have assigned a special role to energy. Our knowledge of the physical world has been acquired from measurements of natural phenomena and of materials, followed by analysis and the recognition of consistent laws, and finally the use of those laws to predict future behavior. If correct results are to be obtained, each step in this process has to be correct. Any error will cause all that follows to be wrong. Therefore it makes sense to carefully understand the measurement process. You will see that energy exchanges *always* accompany every measurement so this process plays a central role. You and I want to understand *fundamental* processes, which probably means we have to look most carefully in the *quantum realm,* where events occur at the smallest level, one-by-one; one particle with another. A lot of these events then make up our view of the other realms.

When we look at measurement methods, we first think of the ways we measure in the *human realm*. There are many methods: we have scales to find our weight, calipers to measure length, comparison charts for color, thermometers for temperature, and so on. There are hundreds of ways to measure. There are so many it doesn't seem as if we are going to find anything basic or a common method that connects them all. The same is true for the *cosmos realm* where we find the same human realm measurements but on a larger scale.

In the *quantum realm* there is less confusion because there are only two types of particles, electrons and protons, involved in most interactions and these affect each other by charge forces; that is, they exchange photon energy. The uncharged particles, mainly neutrons, affect measuring instruments either by gravity, by decay or by collision with other matter. In both cases, the event is revealed by the presence of outgoing charged particles which we can detect by exchanging a photon.

CHAPTER 2/*The Tree of Scientific Knowledge*

Thinking about the process, we are able to suspect that measurement probably always involves exchanging energy by photons. Is this true? When the measurement process is reduced to the basic elements, do our instruments always merely record an energy input? Think about this for a while.

It appears to be true that every measurement is a photon exchange. I can think of no exceptions. This can be examined from another viewpoint. How does an instrument indicate what it is measuring? For example, what causes a meter needle to swing? What magnetizes an iron molecule in a data tape? What moves the liquid column in a thermometer? What blackens the silver chloride molecule on the photographic film? Always, the answer is the same: energy is exchanged. This will turn out to be important, so we conclude:

FACT: *Every measurement requires an energy exchange.*

This is a special role of energy in Nature's scheme. This conclusion has not been universally recognized in the scientific community, so if you agree, you are already a pioneer in the game of logical science.

We can use this idea in our search for truth as it will enable us to examine the conclusions of other scientists, who rarely provide us with their raw experimental data, for making our own deductions. For example, suppose we want to know what really is the meaning of "spin" of a particle. The handbook of Chemistry of Physics says it is, *"Angular momentum which has no counterpart in the human realm."* Use of the measurement conclusion could help us track down the meaning of spin because we know the original data which determined spin must have been an energy exchange. We deduce that no one actually measured angular momentum. Instead, energy was measured and angular momentum was inferred!

What are Force and Power? It turns out that these two quantities are combinations of energy with the other basic units. In order to see how they are combined, we must have a short lesson about some mathematics discovered by Newton, which is calculus.

A Short Lesson in Calculus. One of the neatest branches of mathematics is this useful technique discovered by Newton. It is fascinating because it combines the use of numbers with geometry. Calculus is concerned with rates of change. One of the most important examples is the concept of *speed* which is the *rate of change of length with time*. If length is measured in meters, speed is measured in *meters per second* (m/sec). Speed can change too, and its rate of change is termed *acceleration*, measured in *meters per second per second* (m/sec^2).

The main thing one does in calculus is to find the rates of change of varying quantities. This can be done by algebra or by graphs. For example, speed can be obtained from the graph of the changing position of a moving car in Figure 2-3.

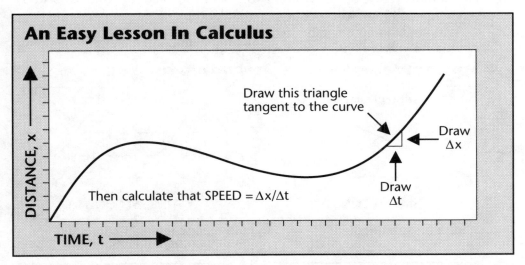

Figure 2-3. Graph of the position of a moving car as a function of time.
Speed is obtained from a graph of distance traveled by drawing small triangles Δx and Δt. The speed is the ratio $\Delta x/\Delta t$, which is also called the slope of the curve.

The rate of change of position is the ratio of the change of **x** to the change of **t**. This is written $\Delta x/\Delta t$. If you want to be very accurate, you make Δt very small and Δx decreases with it. To indicate such smallness the ratio is customarily written dx/dt and called a **derivative**. There! You have learned the basic concept of calculus. There are formulas, easy to learn, found in handbooks for finding derivatives when you already know an algebraic function. Two common formulas are:

1. Formula: If $x = a\, t^n$, then $dx/dt = n\, a\, t^{n-1}$.

Example: a function is: $x = 3\, t^4$. **Then,** $dx/dt = 12\, t^3$.
You put the value of the exponent down in front and decrease the old exponent by one.

2. Formula: If $x = a\, \sin(bt)$, then $dx/dt = ab\, \cos(bt)$.

Example: a function is: $x = 2\sin(4t)$. **Then,** $dx/dt = 8\cos(4t)$.
You put the time factor out in front and change the sine to a cosine.

CHAPTER 2/*The Tree of Scientific Knowledge*

The important concept to learn is that the derivative of a changing quantity is the rate of change with respect to the independent variable. How you find it is useful but not essential to understanding. It is the concept which you must understand fully. Figure 2-4 illustrates two cases of the above formulas to study and check that you have learned the calculus concept.

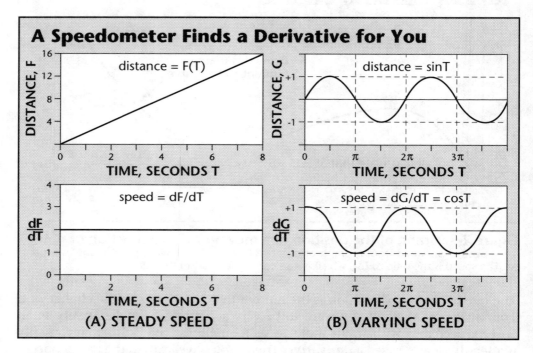

Figure 2-4. Finding the calculus derivative of motion.
If distance F is a function of time, T, then speed is the derivative, F/T of the distance curve. The left graph (A) shows the position of a steadily moving object. Its speed is the slope of the distance curve: shown in the lower graph. The slope, given by the derivative, has a constant value of 2 cm/second. The right graph (B) shows the position of an object moving up and down, like a sine wave, G = sin(T). Its speed, given by the derivative, is a cosine curve shown in the lower graph.

Let's put calculus to work. Using calculus, force can be simply expressed:

DEFINITION: *Force = rate of change of energy with distance = dE/dx*

You can write, force = dE/dx, or, use frequency f as the energy unit, force = hdf/dx. What does this equation mean? Take an example: if two attracting charges were being moved apart, it means that for each distance Δx moved, there is an exchange of a photon whose energy $\Delta E = h\Delta f$, which depends on the frequency change Δf. How the photons move depends on the experimental arrangement. There are four objects in this artificial example — two charges, and two movers, all of which exchange energy, depending on the movements. A force in the human realm is required when a weight-lifter moves barbells from floor to overhead. In this case, Δx is about 2 meters, and ΔE is the energy used for the lift. Note that force is a *derivative* combination of two basic units — energy and length.

In a similar way power is also a combination of two basic units — energy and time. It is defined:

DEFINITION: *Power = rate of change of energy with time = dE/dt*

To calculate the power capability of the weight-lifter you need to know the time required to lift the weights, say $\Delta t = 2$ seconds. Divide that into the energy used ΔE, and you have the power of the weight-lifter: power = $\Delta E/\Delta t$

Fundamental Laws of Nature

Engineers and scientists have hundreds of rules and principles, used for their practice and calculations, which they may regard as fundamental. Luckily for our search of basic understanding, nearly all such rules can be obtained by mathematics from just a few truly fundamental laws. We don't have to be bothered by the many practical rules because we know that they can be derived from the fundamental laws if we ever want them. These derivations can be found in books at the library.

The set of fundamental laws important to pioneers on the forefront of exploration is shown in Table 1. These are the fundamental laws which underlie *all* of the scientific knowledge today. They are the roots of the tree of knowledge out of which all other science grows. In order to move from the root to a branch, or from branch to twig, or branch to branch, all that is required is suitable mathematics and knowing how to use it. Much of the mathematical analysis has already been done, particularly in the region just above the roots where the analysis is uncomplicated.

Of course, there are many remote twigs on the tree which have never been connected by mathematical analysis and the reason is often the complexity of the mathematical work needed. But simple cases have been worked out. For example, the analysis to predict how 2 or 3 billiard balls behave after they strike each other on the table is easy. And perhaps someone has worked it out for 4 or 5 balls too. I don't know. But, I am sure no one has tried to solve the problem for 100 balls. Still, few mathematicians doubt that it could be done if someone provided the manpower, time, and money to pay for it. Why? Because all the rules and the method needed to do it are known.

Table 1. THE FUNDAMENTAL LAWS OF THE UNIVERSE

Law deals with	Common Name	Formula
FORCE LAWS:		
Gravity force between two bodies	Law of gravity	Force = $Gm_1 m_2/R^2$
Electric force between two charges	Coulomb's Law	Force = $q_1 q_2/R^2$
Weak force between two particles	Weak force	Only known to be short range
Strong force between two nucleons	Nuclear force	Only known to be short range
OTHER LAWS:		
Inertia and motion vector	Newton's 2nd Law	$F = ma$ or $F = dp/dt$
Special Relativity (1)	Speed of light	Light speed = c = constant
Special Relativity (2)	Relativistic mass increase	$m = m_o/(1 - v^2/c^2)^{1/2}$
Energy exchanges	Conservation of energy	E = constant, in closed system.
Expansion of the universe	Hubble Constant, $H = 1/T$	$T = 1.3 \times 10^{10}$ years
Quantum waves (1)	DeBroglie wave rule	wavelength = h/mv
Quantum probability (2)	Probability waves, ψ	Probability = $|\psi|^2$

SYMBOLS USED: G = gravitational constant, m = mass, R = distance between objects, q = charge, e = electric constant, v = velocity, p = mv = momentum, a = dp/dt = acceleration, h = Plank's constant, ψ = quantum wave amplitude, F = force vector.

There are some problems which will never be worked out, even though we think we know all the rules. For example, suppose we wanted to compute the properties of an organic molecule in chemistry which has hundreds of thousands of atoms in it. Even if Congress voted to spend a sum equal to the U.S. national debt, it does not appear likely that the properties could be computed. It is too big, but we think it can be done in principle.

Have you noticed that Table 1 has defects? Even though the laws are few, science does not fully know the correct mathematical expressions for all of them. None of the force laws are exactly correct because they have been experimentally found to fail when the distances R become about 10^{-15} meters or less. The Hubble rule has a precision of about 50%. The quantum and relativity rules seem to be in good shape, but you never know when discrepancies will turn up. These problems with fundamental laws will be discussed later.

Fundamental Numbers and Objects. One cannot describe the world of science on the basis of scientific laws alone. In addition there must be fundamental numbers which tell us about the objects in the universe and their properties. The laws operate upon these objects. If there were no objects which obey the laws, the laws themselves would be meaningless.

A list of numerical values describing the fundamental objects is in Table 2. Not all the numbers are well determined, especially those involving the whole universe.

Table 2. FUNDAMENTAL NUMBERS OF THE UNIVERSE (1986 data)

Common name	Symbol	Describes	Numerical value
Speed of light	c	all electromagnetic waves	2.998×10^8 meters per second
Plank's constant	h	basic quantum-unit	6.626×10^{-34} Joule-sec
Charge of electron	e	all charged particles	1.602×10^{-19} coulombs
Hubble's Constant	H	rate of expansion of space	$1/H = 1.3 \times 10^{10}$ years
Electron frequency	f	electron space waves	1.236×10^{19} cycles per second
Mass of electron	m	h/c^2 × electron frequency	9.109×10^{-31} kilogram
Mass of proton	m_p	mass of proton	1836 electron masses
Permittivity of space	ε	wave electric-property	8.854×10^{-12} Coulomb per Newton meter^2
Constant of gravity	G	attraction of masses	6.67×10^{11} m^3 per kg-sec^2
pi meson mass	π^- or π^+	relative mass	273.1 electrons
pi-zero meson mass	π^0	relative mass	264.1 electrons

Frequently Used Combinations:

Fine-structure constant	$\alpha = e^2/4\pi\varepsilon hc$	inverse of 137.04
Electron classical radius	$r_e = e^2/4\pi\varepsilon mc^2$	2.817×10^{-15} meter
Electron Compton wavelength	$\lambda = h/mc$	2.426×10^{-12} meter
Radius of the Universe	$R = c/H$	about 1.5×10^{10} light-years
Density of the Universe	d_u	about 10^{-28} kg/m3
Total particles in Universe	$N = d_u\, 4\pi R^3/3m_p$	about 2×10^{78} protons

In addition to the stable electrons and protons described in the list, there are other short-lived objects, like mu, K, and tau mesons and unstable baryons, which are not listed. The latter have properties like protons but their extra energy (mass-frequency) makes them unstable so they decay into protons, with various lifetimes. None of them have yet been found to have any role in the material world so we won't say much more about them. The combinations listed frequently enter into some of the more famous puzzles of science and we will say a lot more about them.

Drawing the Tree of Knowledge. If I attempted to draw the entire tree of knowledge, I would soon run out of paper, so I have illustrated only the bottom of the tree for the branch of electricity, in Figure 2-5.

It is important to recognize that only a few fundamental laws and objects (shown as the roots in the figure) are needed to obtain everything above. This remarkable fact, illustrating the simplicity of Nature, is one of the things which makes science interesting. Many of the philosophers of science have felt that further simplification of the fundamentals is possible but the means are hidden in Nature, so we must be clever to decipher it. One of the most sought after simplifications, called "Force Unification," is to find a single cause of the four forces in Table 1. At present, causes for none of them are known, but claims are widespread.

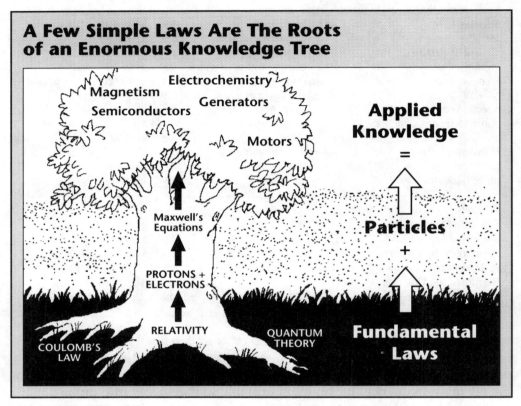

Figure 2-5. The electric branch of the tree of knowledge.
The three roots are relativity, Coulomb's law, and quantum theory. Adding the properties of electrons and protons, one can obtain, without adding further laws or objects, the famous laws of James Clerk Maxwell, and then all the applied rules and principles we recognize as electrical engineering.

 RESEARCH GOAL: *Do All the Forces have a Common Origin?*

When you explore strange realms, it is not easy to deduce the meaning of measurements and obtain the correct conclusions. In the next chapter, we will examine more closely the techniques of distinguishing true from false and fact from fancy.

A reasonable probability is the only certainty.

– Statistician's Maxim

CHAPTER 3

Choosing the Sources of Knowledge

Distortions of Science
Think Quantum
The True Meaning of Measurements

CHAPTER 3
Choosing the Sources of Knowledge

If we want to correctly understand science, we must be very careful to check the entire process of acquiring knowledge from the beginning to the end. Measurement is the beginning. Our last conclusion is the end, and there are many pitfalls in between. Let us a take a short look at the process and the pitfalls.

Distortions of Science

Science in its historical development has frequently been distorted and wrongly conceived. This happens for many reasons. Prominent among them are the social and political pressures placed upon the scientific community by those persons who control the rewards, positions, and promotions which the scientist must have to survive.

Often, the scientists themselves desire engineering, economic, and human convenience in their laws, measures, and concepts. The conclusions we draw from science are always more comfortable to us if they agree with our perspectives viewed from the human realm of space and time.

Although science exists to seek out new knowledge, discoveries which conflict with popular notions are not warmly received by the scientific establishment. Kepler was accused of occultism when he suggested the Moon controls the tides. Copernicus was excommunicated by the church for teaching that the Sun was the center of the Solar System. As recently as 1960, the now-famous author of the Plate Tectonic Theory of Earth crustal movement was ridiculed in his lectures when he stated that South America was once attached to Africa. Only 10 to 20 years ago, the inventor of amorphous semi-conductors (non-crystalline silicon), L. Ovshinsky, was said to be a "crazy uneducated weirdo," partly because he was a self-educated scientist, and his idea was proclaimed "impossible." Now he heads a $100 million dollar NYSE corporation selling amorphous semi-conductors, and is the author of the quotation that begins Chapter 13. On the other hand, not every dissident idea is likely to be correct. As implied by the fabled "Murphy's Law," scientific truth has only a few ways to be expressed correctly, but there are thousands of ways to express it wrongly!

There is an important concept to be learned here and I think it is not sufficient to merely agree that various scientific errors do occur, because you the reader are apt to say, "Oh yes, that happens of course. But it won't happen to me because I am a very careful logical thinker and I am very honest with my own thoughts."

There is the trap! A person hopes and imagines that he is immune if he chooses to be. But the most insidious of scientific errors are those we cannot recognize because: 1) they are ingrained into our cultural thinking, or 2) are a result of our human realm environment, or 3) they have been accepted by the scientific community, or 4) the errors have been knit into the fabric of scientific subject matter.

As the decades pass, science has had to be revised from time to time, correcting errors of historical perspective. The rectification process is usually not just a matter of correcting a number, rephrasing a principle, or adding another term to an equation. Instead, whole concepts may have to be revised, older ways of teaching abandoned, books rewritten, and from time to time, funding by the government painfully taken away from one science organization and given to another.

On a personal level, the process can be quite unpleasant if a new concept forces you to give up a favorite idea or makes obsolete your published book or you lose your research grant. Some scientists are likely to fight the new concept, tooth and nail!

The Beauty and the Beast in Logic and Reason. Beginning in about the sixth century B.C., a period of Greek mathematical flowering, the philosopher-mathematicians with familiar names like Euclid, Archimedes, and Pythagorus, believed that pure human reason was capable of deducing the secrets of Nature. However, history has shown that although their method was immensely productive in the creation of algebra and geometry, their progress in natural philosophy (physics) was very slight. Instead, knowledge of the natural world arose from the work of persons who measured natural events. These persons frequently were architects, engineers, and weapons-makers, like Leonardo Da Vinci, Galileo, and Count Rumford, who sought to understand natural law in order to design buildings and machines. When they were wrong, the machines didn't work, or the buildings fell down. True and false were very apparent, not philosophical.

Formal recognition that science had to be based upon measurement began with Kepler and Newton in the seventeenth century, and became known as the Scientific Method. Nevertheless, human emotions continued to persuade us that pure logic and reason can partially reveal Nature's intentions. A famous example is the work of the philosopher Immanuel Kant (1724-1804), who argued forcefully that time and space possessed independent properties which could be deduced by human thought. His conclusions turned out to contradict the astronomical measurements which led to relativity.

The demise of Kant, and other failures of human intuition in mathematics and physics, led to a tightening of the rules of logic used by mathematicians. Surprisingly, these efforts, which culminated in the work of Kurt Goedel (1906-), led to a strange result: Mathematics cannot prove that mathematics is correct! Goedel showed that the axiomatic method of mathematics has inherent limitations to find truth. He proved that the method of axiomatic arithmetic is incomplete; that is, given any set of axioms, there will always exist true arithmetical statements which cannot be derived from the given set. Further, it is not possible to prove that a system of arithmetic is self-consistent.

We are left with very uncomfortable feelings about our capability to be absolutely sure of anything. In the famous words of Bertrand Russell, "Pure mathematics is the subject in which we do not know what we are talking about, nor whether what we are saying is true." This statement has many interpretations worth thinking about.

One of the more beautiful branches of mathematics is symmetry and group theory. Examples of symmetry in nature, art, and geometry are breathtakingly beautiful. We can stand in awe of the biological mechanisms which grow them, or the skill of the artist, or the complexity of nature. The mathematician is immensely impressed by the wealth of knowledge and application which can arise from a few apparently simple propositions of group theory. Prominent among the applications described by the formalism are the properties of 3 dimensional space. It is very tempting to explore nature with this seductive tool. However history suggests caution.

Think Quantum — Freed From the Human Realm

A powerful bias, unique to 20th century science, which tends to lead us in the wrong direction, is that we live and think in the human realm, whereas the problems we want to investigate lie in the quantum and cosmos realms. Two hundred years of no-quantum, no-cosmos science have spawned attitudes and assumptions of a world occupied by solid particles, euclidian geometry, and preconceived notions of length, time, and mass. How are we to escape this psychological prison?

One way is to be very skeptical and selective about which facts one believes and trusts. Since it is Nature that we are investigating, we must be sure that each fact we use is the actual result of a measurement of Nature. Whenever we examine a particular arena of science, all its background knowledge should be surveyed to check that no extraneous ideas have been added which interfere with the logical path of future exploration.

We must not allow ourselves to extrapolate unreasonably beyond what we have actually measured. For example, protons and electrons both exhibit attractive gravity to each other. Should we give in to temptation and assume that gravity is also attractive for the anti-proton and positron? No. Measurements have not been made and we must simply wait.

We should go back into the garden of cultivated science, and weed out old ideas which are no longer valid on the forefront of investigation. (But don't trash them — they may be still useful for other reasons.) For example, the notions of electric and magnetic fields were created one hundred years ago as a mental crutch to explain action at a distance. They are imaginary things, useful as mathematical symbols and words which make discussion easier, but they may confuse our thinking when we seek the basic causes of electric phenomena. Their actual role should be clarified in our thinking.

The True Meaning of Measurements

In our search for truth we are following the paths of the Geometric and Scientific Methods, recognizing that the only source of knowledge of Nature's secrets is experimental measurement. We must carefully understand the meaning of the measurements so that we can understand Nature. It will never be possible for any one person to repeat all of the experimental measurements that have been made, yet in our work at the frontier it is essential that we be sure of the facts behind us. We know that errors and invalid assumptions continually seep into the substance of science, and a means must be established to find and weed them out. The only feasible method is to read the original accounts of the experimental work which you need to use and apply critical judgement to them. A brief review, here, of some of the more important types of measurements will be useful.

In the human realm, you are already familiar with the kinds of measurements that can be made, so there is no need to describe most of them. The measurements which concern us most are those which affect the quantum realm and the cosmos realm, because we have no ordinary experience to guide our interpretation or check our results.

Quantum Realm Measurements. There are two kinds of measurements from which most of our knowledge has been obtained. These are 1) optical and other spectra, and 2) particle tracks. Examples are shown in figure 3-1 and 3-2.

An optical spectra measurement finds the frequency and intensity of light emitted by atoms which are changing from one energy state to another. The light is passed through a prism or grating which separates one frequency from another. The light intensity can then be seen on a strip of photographic film. What we measure is the energy (frequency) difference between the initial state of the atoms before they radiate, and their final state after radiation. The intensity of the image on the film tells us how frequently this change occurs compared to the other images on the film. It is important to notice that large numbers of identical atoms are changing their state so that large numbers of identical light quanta in each image are recorded on the film. The process is described by the equation,

$$E_1 - E_2 = hf$$

where E_1 and E_2 are initial and final energies, f is the frequency, and the amount of energy change is hf, which is termed one quantum.

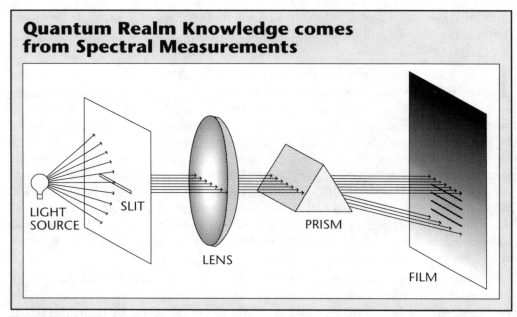

Figure 3-1. Optical spectra.
Often quantum measurements are obtained from the optical spectra of the light emitted from a source which changes from one energy state to another. When frequency (energy) is changed, the difference is radiated by the source. Typically, many identical atoms of the source change their state and their total emitted light is recorded on the film strip.

Spectra are also routinely measured for frequency ranges other than visible light. UV, IR, X-ray, and radio frequencies are involved instead. The main difference is that other types of detectors, rather than film are used to measure the frequency change. Again, large numbers of atoms are involved and large numbers of quanta are recorded.

The various means used to bring about the frequency change are of great interest, for it is often the central feature of the experiment. For example, a magnetic or an electric field can be applied to the atoms, or they may be bombarded with other atoms.

The second means of measurement involves directing single charged particles to enter a medium which will record the particle's path by abstracting part of its energy to create some sort of visible reaction in the medium. Photographic film and vapor-saturated air or liquids are common media. In the latter two cases the passage of the particle causes tiny fog particles or bubbles to appear; hence the method is called a cloud chamber or a bubble chamber. If a magnetic field is present, the particle path is curved and measurement of the path permits calculation of mass, momentum, and energy.

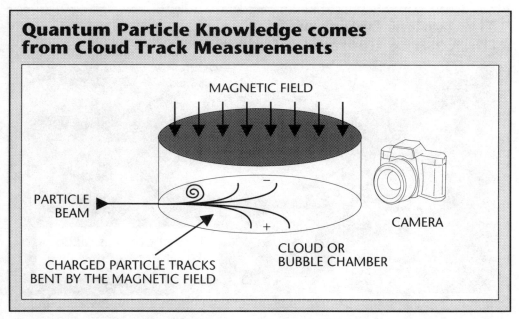

Figure 3-2. A cloud chamber
The gas in the chamber is super saturated and will condense along the path of a moving charged partical. The magnetic field bends the path according to the inverse of the momentum of their particle. A photograph of the tracks is analyzed to find the energy, type, and momentum of the particle. Much information can be gained about the decay modes of short-lived particles and their interactions, since particles are frequently seen changing from one kind to another after the violent collisions which create them.

Each measurement causes an energy exchange, or the equivalent, a frequency change. Other properties, such as momentum or charge, may be deduced by calculation and inference. To make the inference requires added assumptions concerning the relationship between frequency actually measured and the value of the variable deduced. In the human realm, we are confident of these assumptions, but in the quantum realm, there is a degree of uncertainty concerning what is really happening.

Do we know the Laws of the Cosmos? Measurements in the cosmos realm are also records of frequency changes, because radiation from other stars is practically the only means we have of learning about distant regions of space. Cosmos measurements are frequently photos like Figure 3-3. Sometimes film spectra are obtained by the method of Figure 3-1. Motion of the stars can be detected by comparing photos taken at different times. Most measurements are recorded on photographic film, although in the last decade single energy exchanges are measured using very sensitive charge-coupled detectors (CCD).

Figure 3-3. A galaxy of stars similar to our sun.
Careful measurments of the stars in the galaxy provide information about their size, motion, momentum, energy, and the gravity field which holds them together. Measurement of the color, intensity, spectra, and polarization of the starlight yields information about their age and chemical compostion.

Cosmos measurements seldom involve problems of strange quantum behavior, but there is a new uncertainty. We do not know if the laws of physics found on Earth, are also identical in the distant regions of space. This problem has three facets: First, the radiation from anti-matter has been deeply contemplated and it appears that we are unable to determine whether a galaxy is matter or anti-matter. The radiations from either are identical. Second, there is no way to know whether or not the basic constants of physics are the same as Table 2. Within our own solar system, spacecraft measurements have found such constants to be the same. And within our own galaxy, the evidence is strong that anti-matter is rare. Beyond these limits, there is only knowledge, of a general sort, that the laws of physics are at least similar. Third, radiation from the distant regions requires a substantial time to arrive. This means we are unable to know what is happening "now," but only learn what had happened "then" at an earlier time when the radiation first began.

But astronomers and astronauts do have one firm advantage in their work. The cosmos provides us with three dimensional information obtained from the shapes of images on photographs, which is a valuable and different feature of measurement, usually not found in quantum realm work.

"No paradox, no progress."

– Niels Bohr

CHAPTER 4

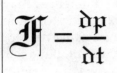

Fundamentals of the Unknown

The Fascinating Tale of Gravity
Every Living Creature Uses "The Conservation of Energy"
Newton's Second Law of Motion
There are Two Special Things About Motion
Conservation of Momentum
Angular Momentum
Conservation of Angular Momentum
The Theory of Special Relativity

CHAPTER 4
Fundamentals of the Unknown

All the objects and laws which lie at the root of the tree of knowledge are the subject of our investigation. In this chapter, we will examine four fundamental laws: Gravity, Newton's Mechanics, Special Relativity, and Conservation of Energy. Because they are fundamental and affect our later analysis and conclusions, we must be careful not to misinterpret nature and not to add concepts which are not really there. Hopefully, our reward may be a glimpse behind the cloud of mystery and a deduction of something new.

After following the careful path given on these pages, you will probably find occasional disagreement with other books and persons. Some differences may be errors of writing. Forgive me. Others arise from the *approximations* of shop talk, customs, and habits, which exist because every discipline has its rules of thumb which, while *not quite accurate,* are satisfactory and very useful. But a few important differences may not be easily resolved. You don't know who is right. Instead, you must remeasure nature or probe deeper with your own analysis. That is what research is all about.

The Fascinating Tale of Gravity

Gravity began with the accurate measurements of a Danish astronomer Tycho Brahe (1546-1601). These measurements were analyzed by the German mathematician Johann Kepler (1600), and finally Isaac Newton (1670) deduced the law of gravity and, equally important, the laws of motion.

Tycho Brahe had heard all the endless religious discussions about heaven and Earth, motion of the spheres, and the quarrel about the Sun or the Earth as the center. He had a new idea, "Why not measure the actual positions and motions of the Sun, Moon, and planets, so that the numbers obtained might settle the arguments?" This was a perceptive and innovative concept which became a harbinger of scientific methods for the next several centuries. He filled voluminous pages with measurements taken at his observatory on the island of Hyen. After his death, his work was studied by Kepler.

Kepler eventually discovered three beautiful rules (See Figure 4-1) about the motion of each planet:

RULE 1: *Planets travel around the Sun in elliptic curves with the Sun at one focus.*

An ellipse possesses an unrelated but interesting property found in "Whispering Caves," that is, if sound is made at one focus point, it will be reflected and refocused at the other focal point.

RULE 2: *The moving radial line from the Sun to the planet sweeps out equal areas in equal times.*

This rule shows that the planets do not move with constant speed but go fastest when they are closest to the Sun and slowest when farthest away. How the rule works can be easily seen, since when the radius is shortest the area is increased most because the line moves faster. Just the opposite is true when the radius is longest.

RULE 3: *The square of the periods of any two planets are proportional to the cubes of the semimajor axes of their respective ellipses.*

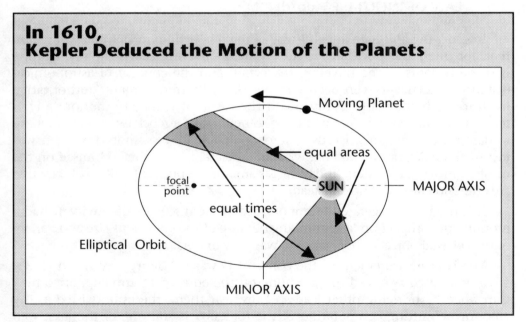

Figure 4-1. The Elliptical path of the planets around the sun.
Study of the motion of planets began our knowledge of the laws of physics. Kepler found the planets move in elliptical orbits with the sun at one focus, and the line from Sun to planet sweeps out equal areas at equal times. Newton realized that the force of gravity, $F = GMm/r^2$ kept them in their paths and that they obey the law $F = dp/dt$.

Which means that the larger the orbital diameter, the longer it takes to get around. The orbital time is proportional to the 3/2 power of the diameter.

Newton's Genius: Newton made the masterful prodigious step. Perhaps because he had studied motion so intensively he appreciated that the Sun was the origin of the forces which governed the motion of the planets. By studying Kepler's second and third law he was able to deduce that the forces decreased as the planets were farther away from the Sun. Eventually, he made the final deduction that all matter has an attraction for other matter proportional to their masses m_1 and m_2, and inversely proportional to the distance r between them. He wrote this in his famous *Principia*:

LAW OF GRAVITY: $F = Gm_1m_2/r^2$

He proved this to himself by calculation and by comparison with Tycho Brahe's observations. In order to do this, he had to know his Second Law to be discussed later:

LAW OF INERTIA: $F = dp/dt$

In order to appreciate the difficulty of these scientific works, you must realize that, for them, the planets were merely specks moving across the sky. No one had the benefits of the traveling spacecraft, and the concept of astronomical distances had never even been whispered. Furthermore, many authoritative institutions had already firmly stated the cause of planetary motion: angels tending crystal spheres. One had to be something of a rebel and a revolutionary to dare think differently. Heads were chopped off for less. Indeed Newton was just such a rebel, but he had the wisdom to keep his thoughts to himself on the most controversial matters. Today, heads are not chopped off, but rebel scientists are often ridiculed and fired instead.

It is hard to exaggerate the importance of the discovery of gravity. It had a monumental impact on later thinking because of the enormous size and variety of the celestial objects whose motion was now understood.

The later measurement of the constant G was by Henry Cavendish (1731-1810) who used two lead spheres hung on the ends of an arm supported by a thin fiber. By bringing other lead spheres near them, the fibre twisted a tiny amount permitting computation of the forces. G is still one of the least well determined constants because it is too weak to measure accurately.

The Scientific Attitudes of the 1600's. It should never be misconstrued that these brilliant men were striving to find scientific laws to be placed in your textbooks of today. Far from it! If you read their original writings you would quickly discover that Tycho Brahe was simply doing his appointed job of getting information that would predict the future and fortunes of his King, using the art of astrology. Tycho believed in such magic and so did most of the intelligent men of his time. His goal was to be the best of the astrologer-astronomers and maintain the favors of the King.

Kepler likewise believed in the existence of heaven, angels, and the teachings of the Church. Why not? Some of the best experts in the land were employed by the Church, so shouldn't they know? Kepler even claimed that he had discovered the "Music of the Spheres." These spheres, made of crystal, were embedded with the Moon, planets and stars, and revolved around the Earth and thus accounted for the motion of the heavenly bodies as in Figure 7-1. According to the theory, the motion of the spheres was accompanied by music. Kepler wrote down his version of the musical themes but no one has played them, probably because the melody is terrible. (Author's note: I tried composing music using Kepler's themes, and concluded that heaven had very little musical harmony.) Kepler, like Tycho Brahe before him, was merely trying to stay ahead of the smart men of his time. He wisely refrained from music.

Newton was a very neurotic man, which was probably the driving force behind his achievements. His deepest instincts were occult, esoteric and semantic and he probably anticipated obtaining magical powers from his work. He had a profound paralyzing fear of exposing his thoughts, beliefs, and discoveries to the inspection and criticism of the world, but he vainly imagined that he had been privileged by God to learn the secrets of Nature. The eventual publication of his works like *The Principia* and *Opticks* resulted from the extreme pressure of his friends, who recognized the value of his work to the world.

His biographers described an important personality characteristic responsible for his achievements. He was able to hold a problem continuously in his mind for days and weeks until it had surrendered to his intuitive analysis and logic. Then, being a supreme mathematician, he would dress it up for exposition. His mathematical proofs were only after-dressing, not the instrument of his discoveries. There is a story of a technical conversation with Halley, who asked Newton,

"Yes, but how do you know that?"

Newton was taken aback. "Why, I've know it for years," he replied. "If you'll give me a few days, I will find a proof of it for you." And in due time he did.

If you are curious to learn more of the inner thoughts of this unique scientist, it is easy to do so, since most of Newton's original unpublished papers are kept at Cambridge University, England.

Question for Thought: *Compare the scientific attitudes of Brahe, Kepler and Newton with the attitudes of scientists of today. Do you think it is possible to guess the future path of productive scientific investigation, and is there a certain way today to distinguish false science from true science?*

No One Understands Gravity. Despite the precise predictions of the equations of gravity when compared to measurements, no one knows the causes of gravity, or understands any connection with any other force or natural law. We know how to use the equation for the force of gravity. We also know the equation looks just like that for electricity except for the constant. We wonder if there is some relation between them? Many attempts have been made to find a connection. If you want to try this, one of the first things you want to find is the ratio of the two forces. Divide the gravity constant by the electrical constant and you get 0.23×10^{-42}, or 0.00000000000000000000000000000000000000000023. This makes you suddenly realize how great the electrical force is and how small gravity is! Why haven't we noticed this before? It is because our experience of the Earth's gravity is created by its mass being much larger than ourselves. On the other hand, we never notice electrical forces from the protons and electrons which constitute most of matter because their forces are perfectly balanced by each other, and the electrical effects we see are due to tiny numbers of unpaired electrons; a small fraction of the number of electrons and protons which constitute matter.

ENIGMA 4-1: *What is the Origin of Gravity?*

If you get a good idea about this enigma, write a paper and try to win the $1,000 annual prize of the Isaac Newton Gravity Foundation in Massachusetts. Anybody can enter.

We have no knowledge at all of gravity in the quantum realm because it is too weak to be revealed in our measurements. But in the cosmos realm, it becomes the central feature of Einstein's general relativity and an enormous amount of mathematical effort has been spent by others peering into nooks and crannies of his equations. However, there is no certain evidence that the general relativity equations actually describe the universe, because data on the distribution of mass in the universe is insufficient, and also simpler ideas can explain the measured evidence available.

Black holes are one sensational prediction of general relativity. A black hole is not actually a hole but a large amount of mass concentrated in one place — enough to prevent light from escaping it because of gravitational attraction between the light's equivalent mass (m'(photon) = hf/c^2) and the hole's mass. Such a place would appear black, since "black" means the absence of light. All light and matter entering the hole would remain there forever. It is presumed that the gravitational and electrical field of the hole would remain detectable. Black holes have not yet been found but have been proposed to exist at the center of certain galaxies to explain an unusual gravity situation.

The conservation of Energy is a backbone of science whose origin is unknown. This great principle was discovered step-by-step, simply by experience. It was a slow historical process taking place over two centuries, because as technology progressed, new forms of energy appeared, each requiring patch-up of the conservation law. As each new form of energy was found, a new formula was added to the equations of conservation. At present, it is thought to be complete and exact; no new patching formulas have been added for forty years. The law states:

Every Living Creature Uses: "The Conservation of Energy"

> **THE CONSERVATION OF ENERGY:** *There is a certain numerical quantity called 'energy' which does not change within a closed system during any natural process taking place within the system.*

A "closed system" means that no energy is put into or taken out of the system.

In a way, the law is very much like the books of an honest accountant; a certain amount of money is put into an enterprise, sums are dispersed for various activities, and then at the end of the year, it is possible to give an accounting of every penny, which matches the starting money. In the case of energy, Nature keeps the honest books for you, but we don't know how she does it.

The idea was very abstract in its earlier years because it was not a description of a mechanism, nor was there any identifiable thing representing the energy. The total energy was simply a number which always remains the same.

One of the reasons the law is so important to us is that a calculable fraction of any store of energy can be turned into useful work, worth money! Work or energy turns the wheels of industry, warms and lights our houses, and makes it possible for some of us to hardly work at all. Work itself is one of the forms of energy:

Work = force x distance moved by the force

Energy comes in many packages. Without a knowledge of the various forms of energy and the formulas which allow you to calculate them, the concept is almost devoid of meaning. Table 3 gives the most common kinds and formulas to compute them.

TABLE 3. Formulas for Energy

NAME	KIND OF ENERGY	FORMULA
Kinetic	Energy of motion due to relative velocity v of a mass m.	$KE = 1/2\ mv^2$
Gravitational	Potential energy of mass m at small height h above Earth.	$PE = mgh$
Gravitational	Potential energy of mass m at any radius r from Earth's center.	$PE = -GmM/r$
Electrical	Potential energy due to a distance r between charges q and q'	$PE = -\mathcal{E}q\ q'/r$
Elastic	Potential energy of elastic material k, stretched a distance d.	$PE = 1/2\ kd^2$
Chemical	Energy of a chemical mass m, obtainable after a defined reaction.	$E = k'm$
Heat	Heat energy of a mass m, due to added temperature ΔT.	$E = k''m\ \Delta T$
Mass	Energy of creation or annihilation of mass m, with anti-mass m.	$E = mc^2$
Photon	Energy of photon of frequency f in an electrical energy exchange.	$E = hf$
Work	Energy given to a system by a force F moving a distance d.	$E = Fd$

M = mass of the Earth. F and d are vectors. The symbols, k, k', k" are empirical constants. Other constants are: G = gravity, $\mathcal{E}$ = electrical, h = Plank's quantum.

You must be careful when using these names since many of them are historical, some descriptive, others practical, etc. In many cases (and possibly all of them, as we will see in later chapters) the differences are due to different perspectives. For example, heat energy given above is a human realm formula, $E = k''m\ \Delta T$, but if examined in the quantum realm, heat is found to be the total kinetic energy of individual moving atoms of mass m, with another formula, $KE = 1/2\ mv^2$.

In recent decades, it has been possible to understand energy somewhat better than folks used to. As a result of the discovery of relativity and quantum theory, we now realize there is an equivalence between energy E, mass m and frequency f, which can be numerically converted to one another by using:

$$E = mc^2 = hf$$

where **h** is Plank's constant and **c** is the velocity of light. Furthermore, we know now how to reduce all of the formulas above for different kinds of energy to mass (or frequency) energy because we know the forces between basic objects ("particles"). This has become possible because the behavior of atoms and molecules in gases, liquids, and solids has been intensively studied. We have tracked the lion to his lair, so to speak. Now all we have to do is find a way to put our head inside the lair and take a look, in order to understand energy itself!

As of this day, we have not been able to get that look, so no one understands the mechanism of the conservation of energy. We only know that the principle holds perfectly in all three realms, human, quantum, and cosmos provided we account for the relativistic increase of mass due to relative velocity.

ENIGMA 4-3. *What is the Origin of the Conservation of Energy?*

Everyone, regardless of education, is familiar with conservation of energy. When you fill up your gas tank, you *trust* that the principle works; you believe that the energy will be there when you need it. Squirrels count on it when they hide nuts for the winter. Insects make use of it when they lay eggs in the body of their prey. This principle is so important, and pervades all of science and life that an explanation of the origin must be ranked very high on our priority list of things to find.

RESEARCH GOAL: *What is energy?*

I will propose an explanation in Chapter 12.

Similarly on the fundamental list, is the second of three laws Newton found using the results of Kepler. The first law was a rephrasing of the *Principle of Inertia*, stated by Galileo Galilei as: *A body in motion continues in that state of motion until it is disturbed.* We can't give Newton the credit for it. However its meaning is also a special part of his second law, which is:

Newton's Second Law of Motion

LAW OF INERTIA: *If a force is applied to a body at rest or in motion, the time-rate of change of its motion is equal to the force applied.*

Newton defined "motion" as mass x velocity. In equation form, his law is

$F = d(mv)/dt$

Today, we speak of his "motion" as *momentum* p. We call the rate of change of velocity the *acceleration* $a = dv/dt$ of the body, so his law is often written, $F = ma$. You can see that this law tells what happens when the momentum is disturbed.

TEST YOURSELF: *Think through the reason why the second law actually includes the first law. Suppose F = 0, then what would the second law say?*

In Science, you are always being bombarded by equations so you tend to treat a new one just like the last. This is a psychological trap; be wary. The equations of motion have special properties which no one completely understands; they are different because they involve the 3D world in which we live. Because motion occurs within the space of our universe, it takes on some of

There are Two Special Things About Motion

CHAPTER 4/*Fundamentals of the Unknown*

the ill-understood properties of space. As a space explorer, you should know that there are mysteries to be unraveled here. Don't miss an opportunity!

If it is Moving, You Always have to Think About Vectors. It is essential to remember that Newton's law is a three dimensional equation describing three 3D *vectors: force, momentum, and acceleration,* which can be drawn like arrows on a paper. Like all vectors, each has three components in three directions at right angles to each other, so there are really three equations, one for each direction:

$$F_x = d(mv_x)/dt \qquad F_y = d(mv_y)/dt \qquad F_z = d(mv_z)/dt$$

Each of these equations is independent and can be solved separately. If motion takes place in a straight line, like a car racing down a strip, you only have to worry about one of the equations because there is no motion in the other two perpendicular directions. In the case of planetary motion, or a rock on a string being swung in a circle, two vectors are involved at the same time and are continually changing. The force vector is along the radius of the string and is a changing combination of, say, **Fx** and **Fy**. The velocity vector is tangential to the circle and the rate of change of velocity is a radial vector perpendicular to velocity. It must be! If it weren't, it couldn't be equal to the radial force.

If you don't have a clear idea of these important vectors, get a piece of paper, draw a circle, and draw a velocity vector at two positions close to each other of an object moving on the circle. Draw the velocity change vector between them. The change of velocity Δv connects the ends of the two velocities. See? It is radial. And Newton's equation tells you the value.

TEST YOURSELF WITH THIS QUESTION: *Is it correct that the velocity of a satellite in a circular orbit is constant, as some people say?*

ANSWER: *No. The speed is constant, but the velocity vector is not. Velocity is constantly changing its direction. The error is the wrong choice of words. A* **velocity** *vector has two parts:* **speed** *and* **direction.** *Speed is the numerical value of the vector and direction tells you where it points.*

Remember position, velocity, and acceleration are always vectors, even if they aren't explicitly stated.

If You Measure Motion, You Must Choose it Relative to Something. Do you remember when discussing the cosmos realm, above, it was pointed out that there is no way to choose a unique coordinate system for velocity or position in the universe? This fact applies to motion of any kind. So, whenever you measure or talk about a length, velocity, or acceleration, you must, knowingly or unknowingly, make a choice of a coordinate system for your measurements. You are free to choose a center for the system anywhere you like: an atom in your finger, your house, the Earth, the Sun, your favorite planet or star.

But don't forget that your choices change the numerical values of position, velocity and acceleration. Consequently, the potential energy, kinetic energy, and forces which depend on these numbers will also change. You can deliberately make choices so that one or more of these becomes zero.

In our earthly affairs, it often happens that a practical coordinate choice is the Earth itself, but even this leads to puzzles because the Earth is rotating and has gravity. Astronomers often use the Sun as a center of coordinates. None relieve us of the confusion that comes from not having fixed values of energy and forces. When people feel confused, they frequently use big words to bolster confidence. This has led to the special words *frame of reference*, to mean a chosen coordinate system. They are used most often when discussing the cosmos realm.

Newton's Law is the Conundrum of the Cosmos. Many have sought to discover where the Second Law comes from. What are its causes and its connections with other laws? There are no answers. No one knows. There is only a hint that it is connected with the entire cosmos. Let's try to see the problem. We are dealing with changes in three basic quantities: mass (frequency), length, and time. We have already seen that force is a combination of mass and length. So, if we knew where those came from we might gain understanding of the Second Law, but we don't. We only suspect they have something to do with space, or maybe the stars. After trying different kinds of acceleration, we also notice that in the Second Law, radial acceleration may occur, so that the velocity and radius vectors may be rotating. How do we measure the rotation? We must measure rotation using the so-called "fixed stars" and assume that the stars are at rest. That is, the stars determine the absolute reference frame of rotary motion. There is no other way to find the rotation!

It is very strange and mysterious that the stars, so far away through empty space, seem to dictate the operation of the most basic law of the universe (Newton's Second). An Austrian physicist-philosopher named Ernst Mach (1838-1916) first noticed and wrote about this. But no one has gone beyond his speculative idea that the distant matter of the universe determines this law of inertia.

Mach's name is now used as a unit of the speed of sound, and a few cars and rock-groups have been named "MACH 1" using his name. His ideas are important so I will discuss him more later. Obviously, the Second Law goes onto the enigma list:

ENIGMA 4-2: *What is the origin of F = d(mv)/dt?*

Newton added a third law which has special uses: *For every force there is an equal and opposite reaction force.* This is commonly written, *Action = Reaction,* but it is not always true, and not fundamental.

His third law is useful in the study of gases or liquids where molecules are constantly bumping into each other, in metals where atoms are tied to each other in lattices, or in nuclear reactions. In these situations you may not know the force laws involved and you may not even have to know, because the Third Law allows you to find overall consequences of force exchanges. The Third Law, together with the conservation of momentum and energy become powerful tools of analysis and logic. For example, using only these few rules you can derive all the properties of compressed or exploding gases in moving cylinders. This is very useful if you are designing engines. You never even have to know the forces between the gas molecules.

There are certain situations involving electromagnetic forces where the Third Law is not valid. However, the Second Law is always true, and if you use them both together, you won't get in trouble easily.

The phrase, "Action equals reaction." is often borrowed by psychologists and politicians who sometimes feel that their forces are comparable to Nature's forces. Is this comparison logical? Can they measure their reaction forces?

Conservation of Momentum

Recall that by conservation one means that things persevere, or "are saved" and not lost. The law of momentum conservation is: *The sum of the separate momentums (vectors) of a given set of bodies remains a constant.* See Figure 4-2.

There is a required condition for this to be true: There are no outside forces to disturb the bodies. In other words, the total momentum at any given time is always equal to the total momentum at any later time. This is true regardless of how the bodies bump and bang each other, but only if there are no disturbing forces from the outside. The classic example is the behavior of the balls on a billiard table. After you hit the assembled resting balls on the table with the cue ball, you can take two quick flash photos, calculate each momentum vector, and add them up to a single vector. Plotting the arrows on a piece of paper, adding head to tail, will work fine. Later on, you take two more quick flash photos, find a new set of different momentums, add them together into one vector, and lo! this total vector will be the same as the earlier vector.

Friction from the table will ruin your experiment if you wait too long between photos, because friction is an outside disturbing force. Similarly, if a ball drops into a pocket the totals won't add up correctly, because you have "lost" part of the momentum.

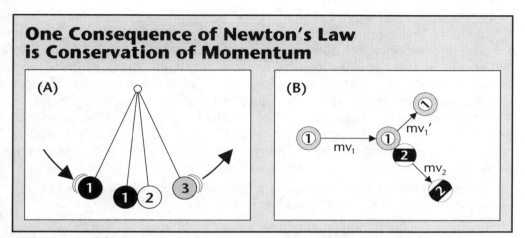

Figure 4-2. Examples of the Conservation of Momentum.
The rule is: "The initial momentum equals the final momentum." In (A), ball 1 swings down and strikes ball 2, which does not move. Instead, it transfers momentum to ball 3, which now moves with exactly the *former* momentum of ball 1, which is now motionless. In (B), billiard ball 1 strikes ball 2 and both move off. The sum of the momentum vectors of balls 1 and 2 must equal the momentum of ball 1 before it struck: $mv_1 = mv_{1'} + mv_2$.

TEST YOURSELF with these three questions:

1) Why is it necessary to take two flash photos in quick succession?

2) How does the total momentum vector of the billiard balls compare with the momentum of the cue ball, before it strikes them?

3) If you really did this experiment and took several photo sets, how would each real total momentum vector really compare with the later total vectors?

There is no enigma concerning the conservation of momentum because it is a direct result of Newton's Laws. We can easily find its origin. The proof is simple: If two balls strike each other, the Third Law and Second Law can be written,

$F_1 = F_2$ and, $d(mv_1)/dt = - d(mv_2)/dt$.

Rearranging, $d(mv_1 + mv_2)/dt = 0$

The last equation means that the time rate of change of the sum of the two momentums is zero, so it is a statement of the conservation of momentum, obtained from Newton's Second Law.

CHAPTER 4/*Fundamentals of the Unknown*   57

The law of conservation of momentum is valid in all three domains: human, quantum, and cosmos. Remember that the mass must be properly expressed by the relativistic mass formula.

Angular Momentum

Similar to linear momentum which we use for motion in a straight line, there is another variety called *angular momentum* when motion is in a circle. Whenever rotation is involved, the concept of angular momentum is most convenient. Knowing that momentum is conserved is a valuable tool for solving mechanical problems, so it comes as no surprise that its operation in three dimensions has been exhaustively studied. From this study, wonderful machines have evolved: gyroscopes, which can accurately guide a vehicle around the Earth; turbines and generators, which provide power to our electric servants; and other rotating engines of all sorts.

Angular momentum **L**, in the case of a single object which is moving in a two-dimensional curve, is shown in Figure 4-3 (these are vectors):

$$L = mx\, dy/dt + my\, dx/dt$$

Study the figure and you will see that **L** is the sum of two angular momentum vectors, one in the **x** direction and the other in the **y** direction. People refer to the lengths **x** and **y** as "lever arms," as if they were imagining the moving mass fixed to a center of rotation.

You can write the angular momentum of the small mass m in circular coordinates, using angle **f** and radius **r**. It becomes:

$$L = mr^2 \times d\theta/dt = m(r \times v)$$

where **r** is the radius from center to mass. You can see that **r** times **dθ/dt** is the linear velocity of the mass in its path. When you use circular coordinates, the mathematics people have discovered the remarkable fact that the two **x** and **y** vectors can be replaced by another vector which is perpendicular to them, and has the magnitude **L** given above. See Figure 4-4.

If, instead of a small moving mass, you have a large rotating object like a wheel, circular coordinates are simpler to work with and Newton's Law becomes

$$L = I\, d\theta/dt$$

where **I**, the moment of inertia plays a role analogous to mass and is computed from $I = M r^2$, and **θ** is the angular coordinate around the center. When you use angular momentum, it usually helps to think of wheels. The moment of inertia which keeps a wheel spinning is the circular equivalent of mass. The

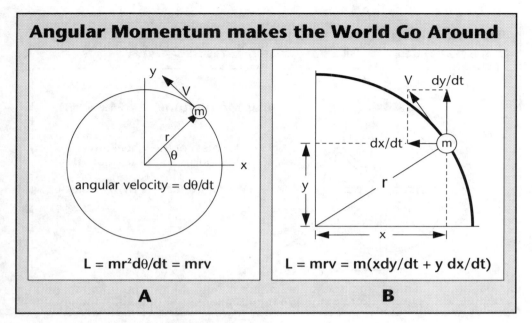

Figure 4-3. A mass moving in a circle
Angular momentum is a property of a mass moving in a curved path. It is a vector which is easiest described in circular coordinates, as $mr^2 d\theta/dt$ as shown in (A). But it can also be described mathematically as the sum of two momentum vectors in the two linear directions x and y, as shown in (B).

angular momentum vector **L** is directed along the axle of the wheel. It is important to recognize that you must choose a coordinate direction for the **L** vector in one of the two directions along the axle, by defining a coordinate system for x-y-z that is either left or right-handed. If you make the usual right-handed choice, the angular momentum of a wheel rotating along the axis of a R-H screw is positive in the direction of movement of the R-H screw, as in Figure 5-1.

Conservation of Angular Momentum

If you have ever repaired a bicycle, you will know that a spinning wheel will continue rotating if there is no friction. This is an example of the human realm conservation of Angular Momentum. This rule is usually stated: *If there are no external torques, angular momentum is constant.* A torque is a combination of a force and a lever arm acting to produce rotation. It has the dimensions of force x radius and is a vector drawn perpendicular to both; the same direction as the angular momentum. The proof of conservation of angular momentum uses the same method as for linear momentum and likewise is a result of Newton's Second law. Similarly, there is no new enigma to worry about.

CHAPTER 4/*Fundamentals of the Unknown*

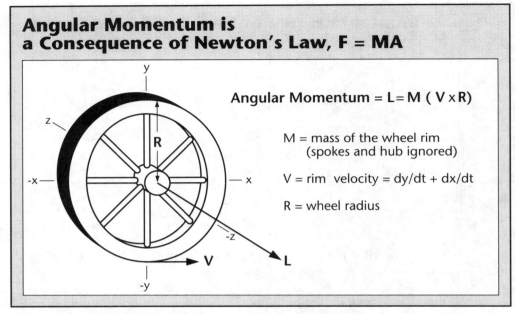

Figure 4-4. How to find angular momentum.
Angular momentum is a vector moving in two dimensions (x and y). The two velocities are vectors dx/dt and dy/dt, but the angular momentum is a vector pointed along the z axis! That is, along the axle of the wheel. The conservation of the angular momerntum law is not fundamental, but comes from Newton's second law, F = ma.

The Second Law can also be written in circular coordinates for any moving object:

The rate of change of the angular momentum about any axis is equal to the external torque around that axis,

dL/dt = external torque = T

This is a vector equation and, because there are three axes in our three-dimensional world, it can also be written as three separate equations. It is remarkable that this law applies to any collection of objects whether or not they are joined together. For example, it applies to the entire Solar System (Sun + planets + moons) whose angular momentum remains the same no matter how many rockets are fired on the Earth. However, one small rocket from another star, not in our Solar System, striking one of our planets will add to the Solar System's momentum. The proof uses the Third Law and assumes that internal action and reaction torques always balance each other out. This is what happens when a rocket is sent out from Earth. It is a powerful theorem which allows us to study over-all motion of an object without worrying about the machinery inside it.

TEST YOURSELF FOR FUN: *Consider a planet rotating around the Sun. Convince yourself that even though the total angular momentum is constant, the two momentum components, say p_x and p_y in the orbital plane, are not constant and in fact are oscillating sinusoids. Since they are not constant, Newton's Second Law says they must be equal to an oscillating force. What is that force? Write out the equation for the x and y components of the gravity force on the planet rotating in its circular orbit. Draw them on a time axis.*

The Theory of Special Relativity

Few things in science have a worse reputation for complexity than The Theory of Special Relativity, but it is completely undeserved. As you will soon realize, the effort needed to learn it is on a par with learning gravity. But you must agree to treat it logically using the experimental facts, and not to be swayed by the fact that you have felt gravity all your life, whereas relativity is mostly on paper. Relativity is also a fundamental law of the universe, like gravity, Newton's Second Law, and the Conservation of Energy.

Relativity can be obtained from two experimental observations:

1) All observers, even though moving with respect to each other, always obtain the same measurement of the velocity of light. The value of **c** is constant.

2) A measurement of mass depends on the velocity of the observer with respect to it:

$$\text{mass} = m_o/(1 - v^2/c^2)^{1/2}$$

where **v** is the relative velocity between observer and the mass, and m_o is the mass when it is at relative rest. There, that's it! The math is no worse than the equation of gravity. Didn't I tell you? But, you say, "I don't understand why these things happen." Well, neither you nor anyone else understands why gravity happens, either. So the two are not different.

Since we rarely do much traveling near the speed of light, the consequences of the theory are strange to us because we haven't experienced them personally. But if you were a designer of TV tubes, the increase of mass would be a routine experience because the electrons which rush down TV tubes do approach the speed of light. If the design didn't take mass increase into account, the picture would be badly focused, and wouldn't fit its space. So experiencing relativity depends on what you do for a living.

CHAPTER 4/*Fundamentals of the Unknown*

QUESTION FOR THOUGHT: *What relative velocity is needed to double the mass? (Ans: 0.866 c)*

Newton's equations, the conservation of momentum, angular momentum, and energy can all be taken over into the cosmos realm simply by using the above relativistic equation for the mass, whenever there is a significantly large velocity involved. Thus the equations for total energy and momentum of a moving object become:

$$E = mc^2 = m_o c^2/(1-v^2/c^2)^{1/2}$$

$$p = mv = v\, m_o/(1-v^2/c^2)^{1/2}$$

A neat result of taking the squares of the above two equations is:

$$E^2 = p^2 c^2 + m_o^2 c^4$$

This relativistic conservation of energy formula shown in figure 4-5, has an easy interpretation. First, the term p^2c^2 depends only on the momentum $p = mv$, so it is like the familiar kinetic energy but in a relativistic form. Then, by adding this kinetic energy p^2c^2 to the rest energy $m_o^2c^4$, you get the total energy E^2. Reasonable. The only unusual feature is: Why must you use the squares of each energy term?

The Only Real Velocity is Relative. Never lose sight of the fact that energy and momentum, both conceptually and numerically, depend on the relative velocity with respect to some other object. This is because no place or object in the universe has been found to be an absolute reference of position. There is no reason to expect there is such a reference. The literature of science in the past did not make much use of this fact. People unthinkingly assumed that the Earth is the absolute reference point for motion, because we live here and overestimate our importance in the Universe. The literature of science still contains definitions, misleading concepts, and ideas derived from this wrong assumption. As a careful explorer, you have to be constantly alert to weed out false assumptions of the past from your thinking.

Another point: We usually speak of energy as "mass" in the cosmos realm, because things are big out there, but don't forget energy is also frequency, $f = mc^2/h$, differing only by a constant of the units which we choose for convenience.

Just as we don't know the reason for gravity, we also don't know the causes of relativistic energy effects. We have another enigma. Chapter 12 proposes an origin for you to consider.

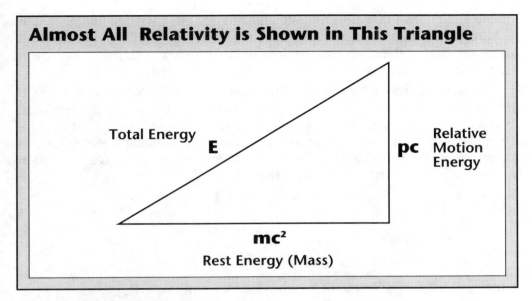

Figure 4-5. The relativistic energy diagram.
The rule says that rest energy plus relative motion energy are combined to obtain total energy. This is quite reasonable, but notice that you don't add them, instead you treat them like sides of a right triangle, and combine them using the pythagorean theorem:
$$E^2 = p^2c^2 + m^2c^4$$
The reason for this method is not understood, but an explanation is suggested in Chapter 12.

ENIGMA 4-4. *What is the Origin of the Relativistic Mass Formula?*

Constant Velocity of Light. Many accurate measurements of the speed of light have been made in several different ways and under different conditions. Despite differing expectations of the scientists, the results are consistent. The verified fact is: If two experimenters measure the speed of light, while moving at different speeds with respect to the source of light, they nevertheless both get the same number.

This experimental fact has interesting consequences which need to be explained. One, it has led to the assumption by scientists that the laws of physics are the same everywhere in the universe, not only in different locations but the same for laboratories moving at different speeds. The truth of this has been amply verified near the Earth, but beyond our Solar System there is scant evidence. This lack of evidence is important. If it is not true it might be that the strength of electric charge or gravity is significantly different in distant galaxies, or perhaps gravity doesn't exist in some faraway place. Objects recently discovered at great distances, pulsars and quasars, emit strange radiation. The apparent enormous energy is mystifying. It may be that we interpret our measurements of them wrongly because the rules may not be the same here as there. We don't know, and we don't know how to find out.

Relativity may also affect Time and Length. Another consequence is that time and length appear not to be constant. It can be quickly shown that if two differently moving observers obtain the same value for the speed of light, then they should have differing clock rates and differing length scales. Each of them figures his time intervals are longer than the other observer, and each calculates his own lengths are shorter than the others.

The following two formulas are obtained for two observers who are moving with speed difference v:

own length = other's length x γ

own time = other's time/γ

where γ is the same factor as in the above mass formula: $\gamma = 1/(1-v^2/c^2)^{1/2}$.

Are these formulas a fact or a paradox? Their correctness has never been verified by direct measurement, since it is very difficult to do, but both have been indirectly presumed to occur by measuring other things. This has led to controversy. Few quarrels have occurred over the length change, but the time change is immediately challenged because of a "Twins Paradox" seen as follows:

The Twins Paradox of Relativity. A space traveler sets out on a journey at relativistic speed leaving his twin brother back on Earth. When he returns, he calculates that his brother will be younger than him. But his brother also makes the calculation and decides his traveling brother is younger! Impossible? Many arguments have ensued over this paradox and many explanations have been proposed which depend on acceleration while turning around, starting or stopping, or passing through Earth gravity. These are arguments of asymmetry.

The situation can be made symmetrical. Assume two twin space voyagers set out to travel at high relativistic speeds in opposite directions. Both go the same distance, turn around, and come home. Then according to the formulas each expects his twin brother to be younger than himself because of the speed difference between them during their travels. This is obviously impossible.

Most physicists take sides on this paradox. I have sought out the writings of many Nobel Prize physicists and it is interesting that they each explain the paradox differently, although all of them firmly agree, as I do, that other aspects of special relativity are correct. Since no crisis in physics would be created by failure of the time formula, no one panics. The future resolution of the paradox could go either way, or be explained by another interpretation.

Relativity and QM Rule the Universe. The important substance of relativity is the change of mass and the conservation of energy which together with Newton's "F = ma" law form the basis of mechanical science. It is thus frequently said that relativity and quantum mechanics together rule the universe. Indeed they do, and most of us would know this if it were not that, living in the human realm, we rarely encounter them with direct experience.

This chapter completes the basics of science as known before about the year 1910, and includes most of our knowledge of the human realm which has been little changed since. The chapters ahead will include the two scientific revolutions which moved quickly in the 1920's: quantum and cosmos. Knowledge of the quantum realm was led by the theory of quantum waves and experimental details were added mostly by particle accelerating machines. Knowledge of the cosmos realm began with telescopic observations beyond our galaxy and was catalyzed by Edwin Hubble's theory of the expanding universe.

"A fish probably has no means of apprehending the existence of water; it is too deeply immersed in it."

– Sir Oliver Lodge

Chapter 5

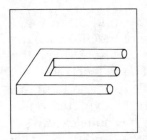

On the Importance of Living in Three Dimensions

The Politics of Living in Lineland
The Aesthetics of Living in Flatland
Symmetry, so Simple, so Useful, We Never Notice It
Left-hand, Right-hand, Upside, Downside, Round & Round
What is Rotation in Space?
The Speedy Photon in the Slow-Motion World

Chapter 5
On the Importance of Living in Three Dimensions

Ideally, scientific truth must result from careful observation of Nature and be followed by logical and mathematical analysis. Truth, however, does not always make her appearance so easily. Fact and fancy, truth and falsity seem to be forever entwined in the fabric of society and in the properties of the space in which we live. I offer the following two allegories from other worlds to aid in comparisons with our world. Then I will discuss several interesting properties of our 3D world.

The Politics of Living in Lineland

There is a story about strange creatures who live in a world of only one dimension; their universe is comprised of a single line of indefinite length. Their environment consists of other creatures and objects whose shape is a point or a piece of a line. These beings live in "Lineland." Life in Lineland is not very eventful because there is not much to do except shuffle back and forth in a monotonous way between their two neighbors. As might be expected under these crowded circumstances, they also do a lot of talking. It is quite exciting to see how far a message can travel from one neighbor to another, and then finally get back to you. Passionate friendships are often made between distant Linelanders who never meet each other end-to-end, so to speak.

Linelanders, like all creatures who live in urban surroundings, govern themselves with a constitution that guarantees them certain rights. One of these is Freedom of Space; no Linelander must ever be deprived of space to shuffle. Just to think of the possibility is so unpleasant that discussing the subject is considered obscene. "No-space to you!" is an insulting epithet. Another right is the Freedom of Speech. For them, it is important that *all* messages be transferred correctly from one neighbor to the next without delay or interruption. Without this, the fundamental act of speaking would be meaningless.

Some Linelanders are geometers who kindly devote themselves to the study of lines and points for the benefit of their fellow creatures. Some of them speculate about a world of zero dimensions, called "Pointland", whose inhabitants must be very crowded, all living at a point. The geometers feel very grateful that they live in one dimension, and during emotional moments, sneer at Pointlanders with an air of superiority.

In one Lineland kingdom, a group of geometers have discovered that if a message is transmitted continuously in one direction, a similar message eventually returns to the author from the other direction. Using this discovery they proclaimed that Lineland exists in a "circular" universe which is embedded in another space of two dimensions. Most simple Linelanders regard the concept of a "circle" as fantastic. They point out that messages sent by others have never returned and therefore Lineland is straight and extends to infinity. But these statements do not deter the geometers, who are inclined to vanity and claim that logic is on their side.

Many inflamed messages have passed through the kingdoms of Lineland taking one side or the other other of this controversy, but nothing has changed. Of course, in Lineland nothing much can change.

The Aesthetics of Living in Flatland

The inhabitants of the two-dimensional world of Flatland enjoy life and its activities carried out on a flat plane surface, in a fashion not too different from human life on the surface of a sphere. Our third dimension of height is not recognized or missed and, therefore, creates no difficulties for Flatlanders.

The creatures of Flatland view their countries with complete satisfaction, being very nationalistic, and have highly segmented societies. Citizens are differentiated by the figures of plane geometry. Authorities recognize the divine spiritual beauty of symmetry, and accordingly the members of society which possess circular shapes - a totally symmetrical figure - have the highest status. Lesser beings have forms like regular polygons and occupy positions in the social hierarchy according to the number of their sides; a triangle being at the bottom, a square is higher, a pentagon above that, and so on. Deformity is unforgiven, so that those who have no symmetry become the serfs and menials of Flatland, a position regarded as logical and suitable for citizens who lack the barest essentials of geometrical harmony.

The thoughts and ideas of the geometers, circular citizens of course, are highly regarded. So that, one day, when a lecturer said the sacred number "Pi" (3.14159787) was not constant, as stated by the authorities, a great conflict arose between geometers, some taking the position of the authorities. It was claimed that Pi varied, depending on the size of the circle measured.

Measurements were made in one kingdom, named Dome. If the circle used to measure Pi was made larger and larger, the value of Pi became smaller and smaller. The dissident geometers then created a concept called "space" which had three dimensions. Then they argued that Flatland was actually a two-dimensional surface which formed part of a "sphere." (The sphere was a 3D object created by rotating a circle around its diameter, into the third dimension). They pointed out that Pi on a sphere is smaller. This idea made a lot of sense to some Flatlanders who saw that the sphere was an extension of the perfection of a circle and logically could be true.

Similar measurements were also made in another Kingdom far away, named Wrinkleton. But here, the value of Pi became larger when the measuring circle was enlarged. In Wrinkleton, geometers also formed the idea of three-dimensional space, but instead claimed that flatland was a two-dimensional surface which formed part of a "wripple" (a wripple is a 3D object created by expanding a circle, that has a rippled circumference, outward from its center into three dimensions). On such a surface Pi is larger. This idea was not at all popular because a wripple could not have the perfection of a circle or even a sphere. Quite the contrary, it was deformed! Clearly it could not be part of a divine natural plan.

The resulting quarrels and battles which take place amongst the Flatlanders have never been resolved. They are still attempting to grasp the problem.

A long story about creatures who live in a universe of two dimensions was written (about 1896) by Rev. Edwin A. Abbott. His book "Flatland" described the problems his Flatlanders had in accepting new ideas; how to envision three dimensions when you live in two, plus other social issues. It later became a prophecy concerning the confusion created by Einstein's Relativity, twenty years later. Today, we are still trying to grasp the problem.

I have told these two stories for two reasons. One, to illustrate how common it is to base scientific argument on social issues. Although it is easy for us to see the flaws in Flatland, it is not so easy if you are a member of the society. This problem exists in our society and untangling truth is quite difficult, if you do not already know the answers. The second is to introduce the ideas of 1D space, 2D space, and 3D space, because Einstein's disciples have made use of 4D space in quite the same way as the Flatland geometers used 3D space.

Beyond 4D space, there are respected geometers of our modern society who suggest there is a 5D space, a 10D space and an 11D space. The concept of symmetry is also given a special status in today's science. Are these ideas real? Or are they flawed arguments? You must decide.

Symmetry, so Simple, so Useful, We Never Notice It

Symmetry is an aspect of geometry which is very helpful to understand the physical world around us. It makes use of the symmetry we find in Nature. It is often geometrical.

If you ever listen to some lectures on group theory and abstract spaces, you can easily conclude that symmetry was brought into science with the intention of confusing people. But you can demand, "Just a minute, symmetry usually involves a simplification of construction or design. Shouldn't it therefore allow us to see the mathematics and science more simply?" Of course. That is the way it is. You had just wandered into a difficult lecture. As your advisor in exploration, I can tell you, "Whenever you see a situation of symmetry in a physical problem, stop and think! Because you will nearly always find an easier way to solve the problem by using the symmetry property." This is one of the rewards of playing around with symmetry. The ideas are neat.

Most often we think of symmetry in terms of art and design. When we see a painting or a sculpture that has one part just like another part, we say it is symmetrical which is a word from the Greek meaning *equal*. The equality of artistic form may refer to parts on either side of a plane, on either side of a line in a drawing, portions around a point in a solid or in a drawing, or sometimes around an axis. Clearly, artistic symmetry can be one, two or three-dimensional and is related to geometry. Part of the beauty is often related to the simplification.

In mathematics and geometry, there is a need to be precise; so there symmetry is defined to mean that a function or a geometric figure remains the same, despite: 1) a rotation of coordinates, 2) movement along an axis, or 3) an interchange of variables.

In physical science, which is our main concern, the existence of a symmetry usually means that a law of Nature does not change, despite: 1) a rotation of coordinates in space, 2) movement along an axis through space, 3) changing the past into the future such that t becomes -t, 4) an interchange of two coordinates such as exchanging x with y, z with -z, etc. or, 5) the change of any given variable.

An enormous number of symmetries exist which are of no interest to us. For example, Nature is symmetrical with respect to politics. She couldn't care less whom we elect, and her basic laws are unconcerned with the number of chickens in the universe, and so on. But notice that if an *asymmetry* should appear (the symmetry is '*broken*' in math lingo), no matter how strange the variable is (even chickens), if there is a dependence of natural law on that variable, it would suddenly become of vital interest to science. An example of a symmetry is if the gravity force changed with temperature. So in science both symmetry and asymmetry are important, depending on the situation.

How Symmetry Helps Solve Problems. Often symmetry can allow you to solve problems with very little math. For example, when you throw a stone into the air, you observe the laws of physics are no different when it goes up compared to when it goes down, and they are the same to the left and right of the top of the trajectory. As a result you conclude the shape of the path is the same going up as going down. The math becomes much easier. Another example occurs if you are installing radio receiver and transmitter stations. There is a symmetry for propagation of the radio signal – it travels in the same manner no matter which direction it goes. So all you need do is calculate the power you need for one direction; the other is the same.

Tiny asymmetries are often quickly recognized because the effects they cause stand out even though they are small. For example, Newton's laws have no dependence on relative velocity, so they are written symmetrically with respect to velocity of a particle. This symmetry is wrong, both logically and according to our careful observations, because it involves assuming that the Earth is at rest. After we study astronomy we find there is no object in the universe at rest. So in fact the old laws must have an asymmetry when relative velocity occurs.

The Symmetry of Special Relativity. According to Relativity, which is the logical way of thinking, only the relative velocity vector between two objects is absolute. In most situations the tiny variations it produces in physical laws can be calculated using Newton's Laws with an additional term containing the tiny factor v^2/c^2. This calculation treats relativity as a cause of asymmetry, or 'breaking of symmetry.'

When you see small asymmetries, you have a built-in method of easy calculation. Just use the prior symmetrical law, and add one (tiny) correction factor containing the new variable. That's all. You will always recognize a relativistic equation because v^2/c^2 is always there. This viewpoint in the case of relativity was not used historically because the symmetry-asymmetry idea has only been developed recently.

All the rules of thumb which have been painstakingly measured during the last century for calculating magnetic effects, contain a relative velocity factor v^2/c^2, plus the Coulomb electric factor e^2/r^2 in varying geometries. This was never noticed before about 1960, even though relativity began in 1905. What does this suggest to you? It says magnetism can be regarded as an asymmetry of the electric laws, caused by relativity. This situation, too, was not recognized historically. Even now, few persons are aware of it. You are among the pioneers.

There is also a symmetry in Relativity, itself. Given any two particles of the same kind, which are interacting together in any way; whatever is true for one must be true for the other. The reason is that the relative velocity between them is the same for both. Each must experience the same force from the other.

QUESTION FOR THOUGHT: *Do you think magnetism is really only a relativistic effect of the more basic laws of electric charge? If you can convince yourself this is so, the idea of magnetic charges or poles must be incorrect. Therefore, magnetic poles do not exist. Many scientists today believe they do exist, although none have ever been found. What is the truth?*

ANOTHER QUESTION: *Another interesting situation, potentially important and yet unsolved, is the gravity force which is tiny compared to electric charge force. The rule for its calculation is exactly the same as the electric force, except for a + sign for gravity, and the value of the constant which is 10^{-42} times smaller than the electric constant. Is there some variable of the Universe, which possibly has a broken symmetry, which is causing a tiny disturbance of electric charge, which we call gravity? This possibility has never been proposed before, so you are one of the first persons to read and think about it. I will discuss it in later chapters.*

Left-hand, Right-hand, Upside, Downside, Round & Round

Everyone has noticed the way a mirror changes our right-hand to our left-hand, or has been frustrated when turning a left-hand screw that you thought was right-handed. This duality of rotation in space is a unique property of our three-dimensional world. This special property allows us to use a vector for angular momentum or torque that points in a direction perpendicular to the two vectors which produced the torque! Mathematicians have created a special notation to represent that feature of two vectors. It is to write torque **T** as a "cross-product" of the force **F** and the lever arm **R**, shown in Figure 5-1, as, **T = F x R**.

The cross **x** operation means that the vector F is turned toward the vector **R** as though you were twisting a screw. Then the direction of **T** is defined as the direction that a right-hand screw would move. You suspect right away that there is something different about the vector **T** compared to the ordinary *"polar"* vectors **F** and **R**. It is because you must agree on whether you are using left-hand or right hand space before you can write the vector **T**. For this reason the result of a cross-product is called an *axial vector*.

Strange things happen when nature is involved with cross-products and axial vectors. In the human realm, the unusual behavior of the gyroscope or "top" is a fine example. Why doesn't it fall over? Another example, involved with the weather, is that the Earth's surface temperature tends to force air from pole to equator but the rotating Earth (axial vector) creates a "coriolis force"

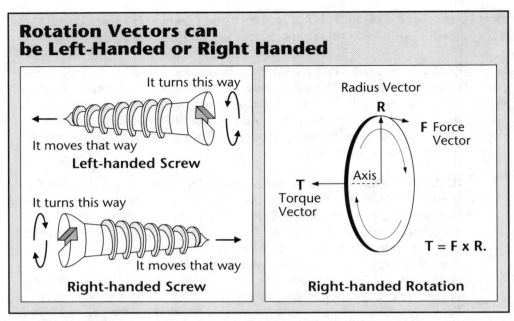

Figure 5-1. Rotation is a special property of 3D space.
The geometry of 3D space has two ways to describe rotation which we call left-handed (L-h) and right-handed (R-h). They are distinctly different. For example, an L-h sugar molecule cannot be absorbed by a living cell which expects an R-h molecule. L-h and R-h electrons have opposite charge and force properties. Some consequences are deep and puzzling.

which causes the wind to blow from West to East in the Northern Hemisphere and East to West in the Southern Hemisphere. For the same reason, cyclones twist clockwise in the Southern Hemisphere and counter clockwise in the Northern. If we did not live amongst these weird forces, we would have great difficulty understanding them at all.

Knowing this, we can expect doubly strange behavior in the quantum realm, where we cannot have personal experience. A prime example is the electron's "spin." No one understands spin yet. The electron is spherically symmetrical, but the spin seems to behave like an axial vector! How can there be a vector in a totally symmetrical situation? Impossible? We will see in a later chapter how the electron's spin is inextricably woven into the properties of 3D space in a fascinating way. It is a great clue to a puzzle that is just waiting to be solved.

The Imaginary Exponent, i. For years mathematicians were puzzled by the meaning of the operation of taking the square root of a negative number. It is still puzzling, but they made sense out of it after it was discovered 100 years ago that the following equation is true:

$$e^{i\theta} = \cos\theta + i\sin\theta \quad \text{where } i = (-1)^{1/2} \text{ and } \theta \text{ is an angle.}$$

This is a remarkable result that has many uses.

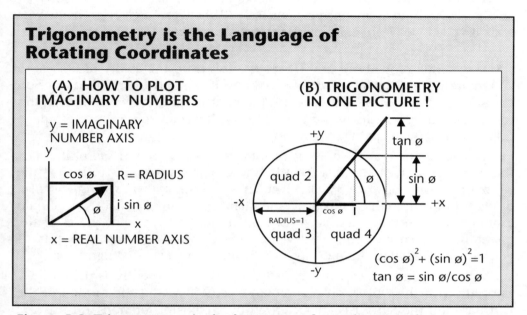

Figure 5-2. Trigonometry is the language of rotating coordinates.
(A) Imaginary numbers are really a convenient way of handling two numbers at a time. The y axis is defined to be the axis of imaginary numbers. All y numbers are multiplied by the square root of -1, or "i." It is a two number system separating x's from y's. From trigonometry, the radial vector R is equal to the cos ø + i sin ø.
(B) shows how all the basic trigonometric relations can be drawn in one diagram. The sign of any quantity in any quadrant is found by using the + or - sign of its x or y values, and computing the sign according to the usual rules of division. Apply the pythagorean theorem to the triangle and get the basic trigonometric algebra. See diagram.

This equation can also be expressed geometrically by the diagram of Figure 5-2 (A). You see that **i** can be helpful in changing from rectangular to circular coordinate systems. Diagram (A) can be expanded into another like 5-2 (B). Diagram (B) is very interesting because it contains all of the basic information one learns in a trigonometry course. Memorize it and try the games that follow.

Algebra games:

1) Surprise your friends by pointing out that the four algebraic symbols π, i, e, -1 can all be squeezed into one expression: $e^{\pi i} = -1$. Suppose you wanted to have a zero in it, what can you write?

2) Find a book that tabulates series for algebraic functions. Write the series for the sine, cosine, and the exponential functions. Put i and θ from the above formula into each term and you will see clearly why the formula is true.

3) Use the diagram (B) to find the standard relation: $\sin^2 x + \cos^2 x = 1$. What are the signs of the cosine in quadrant II, IV? Where can a cosecant be drawn? Neat, isn't it? If you learn diagram (B) you will never forget trigonometry!

What is Rotation in Space?

A very subtle property of our 3D space arises from the ability of objects to rotate. The concept of rotation as we know it can only exist within three dimensions of space. Most of us take this for granted, but if we were 2D creatures we would not have this important property. It makes some sense to take a deeper look at the meaning of rotation.

Strangely, there are two kinds of rotation, *cylindrical* and *spherical*; a fact that is not at all well known except by the geometers who like to look at the curious properties of space. The kind which is not well known, *spherical rotation*, displays properties which link it to the quantum realm. Therefore we should examine the two kinds and keep them in mind; perhaps some pieces of the quantum-space-cosmos puzzle will fall into place when you least suspect it. (If you suspect it will happen in Chapters 7, 8, and 13, you guessed right!)

Most everyone knows about *cylindrical rotation* because this is the kind we associate with the wheels on our car, an electric fan, and the motion of the Earth around the Sun. It is characterized by an axis of rotation and it obeys Newton's usual law of inertia $F = ma$, in cylindrical coordinates. The rotating body possesses kinetic energy of motion, just like any moving body. If you reverse the *axis* of motion end for end, that is, change a coordinate, say z, from z to -z, it is equivalent to reversing the direction of rotation.

Spherical rotation is different from cylindrical. It takes place around a point rather than an axis. The rotary motion, if we apply it to space rather than physical objects, does not result in the accumulation of energy since only space, not matter, is rotating. Also, the motion preserves the continuity of space. If you tie a rope to the spherically rotating body, the rope will *not* wind itself up and break!

This notion already sounds so strange that you are probably wondering if you should believe it. To alleviate your doubt, I will allow you to prove the motion to yourself by constructing an object which will do these things, as in Figure 5-3. You can make the object with not just one rope but six ropes; in fact, you could use any number.

Figure 5-3 is easy to make using a soft cork and six long rubber bands which stretch easily. You need to find some kind of a wooden framework, shaped like

A Home-made Model of Spherical Rotation

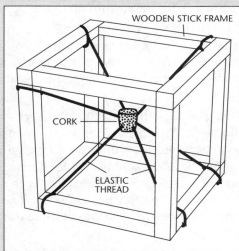

This apparatus is easily made out of a few sticks, a cork and six rubber bands. The cork can be rotated (taking care not to knot up the rubber bands) continuously without entangling the rubber bands!

The cork and bands will return to their initial configuration every two turns.

This serves to demonstrate a little known variety of rotation. It has application to particle theory because the spherical rotation does not destroy continuity of space.

Figure 5-3. Rotation which preserves space.
A remarkable property of 3D space is that objects can rotate in a "spherical" fashion which does not twist up the coordinate lines of space. In this model the rubber bands represent coordinate lines. Do they twist up? Try it.

an open sided box with rigid sticks forming the edges. You could also use a strong cardboard box with one open side so you can get at it. Fasten the elastic bands to the cork with a stapler or a needle piercing the cork. Fasten the other ends to the sides of the box attaching them anywhere. The six bands can be regarded as three coordinate axes.

Try rotating the cork keeping count of where you are. One rotation of the cork will twist the bands. Now, you will discover to your amazement that you can rotate the ball with its attached strings *two* turns (720°) after which, the ball and threads will be back where they started! Amazing! This is hard to believe — I didn't trust it either until I tried it. To avoid turning problems, first choose the main direction you wish to rotate, then turn the cork sideways about a half turn as you go along to avoid tangles.

Suppose the Cork were a Particle in Space? The reason I am discussing this strange rotation is because it might actually be the behavior of a fundamental particle. If so, we may not be involved with a solid human realm object. If applied to an electron which was constructed out of space there is no object there at all, only the lines of space are distorted. Our cork model is only a useful device to see what happens to the space lines as we rotate the cork. Since this type of rotation is not a wheel or other object, we must abandon the notion of momentum and a rigid object, and keep only the coordinate lines of space in our thinking.

Does Space Have a Property of Continuity? What does this rotation property imply for a particle? It suggests that if the cork represented the center of a fundamental particle, and if the particle were constructed of the fabric of space, then spherical rotation makes it possible for the particle region to rotate and yet maintain continuity with the rest of space.

No one knows if space really behaves like this, but many of the great natural philosophers including W. K. Clifford, Sir Arthur Eddington, Hermann Weyl, and Albert Einstein, sought for a geometric property of space to explain matter. They realized that if the matter (particles) are actually distortions of space, the motion of the matter must not destroy the *continuity* of space. At that time none of them knew about spherical rotation, which has been recently found. Now you the reader are a pioneer with a few extra ideas in your bag that they didn't have!

If space has continuity such that its coordinate lines are always maintained, then restraints are imposed upon the motion of spinning. It is not allowed that the lines be stretched without limit, as would occur if a particle underwent cylindrical motion. Further, motion cannot take place that would have one part of space sliding past another since again the lines would not maintain continuity.

Group Theory Describes Spherical Rotation. Spherical rotation can be represented by a mathematical "Group" which makes it easier to look deeper into its geometric properties and to understand and predict what happens in a real physical situation. It is not necessary for you and I to fully understand group theory. Instead we can trust the mathematical conclusions of those who study it, since their conclusions and proofs are open to scrutiny by other mathematicians who are always eager to pounce upon errors to keep them honest.

Group theory is a branch of mathematics which deals with operations that can be performed upon objects of varied geometric shape and varied dimensions. Groups are collections of similar objects, points in space, geometric operations or other things upon which mathematical operations are performed. Group theory can describe the properties of space; so it becomes a useful tool of analysis for us. Our application must be rotation that takes place around a point; that is, the size of the object must be of no concern. This is because group theory is not involved with physical dimensions — it analyzes other properties of geometrical objects.

The requirement of *continuity* has been shown by the mathematicians of Group Theory to be equivalent, in their shop-talk, to a requirement that the motions be represented by a *simply-connected* group. This is a very elementary property whose meaning is clear by noting that a sphere is *simply-connected* whereas a doughnut (with a hole) is *doubly-connected*, a doughnut with two holes is *triply-connected*...and so on.

A Close Look at The SU(2) Group. Spherical rotation can be represented by a mathematical group *space* called SU(2). The abbreviation SU(2) means: *Special Unitary Group of Order Two*. It is a *simply-connected* group, and *Second Order* means that the members of the group are composed of two elements, like a complex number. *Unitary* means that the length of the members (the members can be vectors) in the group space have unit length (|V|=1).

A complete description of spherical rotation can be represented by SU(2) if the rotation takes place around a point and the dimensions of the object are of no concern. For this reason, the question of rotational energy is also not involved. The mechanical model demonstrates the important properties about the group, without much of the mathematics. So don't worry about details, unless of course you discover you like this kind of math. The following is a brief description of how group theory can represent the spherical rotation geometry.

First, what do we need to represent? A spherical rotation is defined by 1) stating an axis of rotation, L, and 2) stating the amount of rotation, an angle Ø.

A group space for this spherical rotation can be chosen to occupy the interior of a sphere. Choose an origin, say point "O" for the spherical rotation at the center of the sphere. Represent the angle of rotation Ø, as the distance away from the center. Obviously there are many positions which represent a given angle. Then choose a diameter of the sphere to represent the rotation axis L. Each value of rotation Ø is d units away from the center on the diameter, and a counter rotation of angle -Ø, is -d units in the opposite direction.

The sphere is large enough to represent all possible positions if the diameter of the sphere is π units. This is because the other half of the sphere represents, by counter rotation, up to -π units. A complete 2π rotation is thus represented. Our complete group space occupies a sphere of diameter 2π angle, and every point in it represents a particular rotation around a particular axis. Notice that the two end points of a diameter represent the *same* point. This group space has been given a name, "the orthogonal group in 3 dimensions" and a symbol, O(3)+ and with a little modification, below, it will become the SU(2) group of interest to us.

Group Properties. The character of any group depends upon its geometric properties: 1) An important group property is that any path between any two given points can be deformed into any other path between those two given points. This is called *homotopic*, and provides us with an important property for particles waves which is that the continuity of (real) space of the universe is not violated. 2) Another property refers to *closed paths*, that is, paths which begin and end at the same point. If every closed path can be shrunk down to a point, the space is termed *simply-connected*. For example, a closed path around the periphery of a doughnut cannot be shrunk to a point because the hole in the doughnut gets in the way! In general, such a hole makes the space *doubly-connected*, three holes make it triple, etc.

Is the O(3)+ space simply-connected? No, because any path which reaches the surface of the sphere leaves the space and leaps to the opposite side. It becomes impossible to shrink the path since the two points are always diametrically opposite. So, the O(3)+ space is not suitable for us, but there is a way to get out of this fix! It will lead us to SU(2).

Suppose there were two O(3)+ spheres and a path which left the first sphere entered the second sphere, crossed it, went back to the first sphere, and returned to the starting point. Now any such path can be shrunk to a point, which will lie on the surface. Great. We have gotten out of the first fix, but now we are in another — we have two spaces instead of one! But, there is a way to get out of this, too.

SU(2) — Finally. We can devise a geometry in which the two spheres become one: We can join the two 3D spheres to become a 3D surface of a 4D *hypersphere*. To see how this is done, first consider two disks with paths entering and leaving at diametrically opposite points. Each of these disks can be deformed (beat them in the middle) to become hemispheres. Now, those two hemispheres can be joined so that the entering and leaving points match up. Thus, the two 2D disks have become one spherical surface of a 3D sphere, and they are *simply-connected*. Here is the neat trick. The same joining operation can be performed with two 3D spheres to become the 3D surface of a 4D hypersphere. That space is SU(2). Simple, isn't it?

How big is this hypersphere? Recall that one turn (2Π angle) of the cork in the spherical rotation model was equivalent to crossing the diameter of one O(3)+ sphere. In contrast, doing a complete circuit of SU(2) corresponds to a double turn of the cork and goes through 4Π units of angle. On the surface of the hypersphere, this is one circuit around its circumference. So the circumference of the hypersphere is a distance of 4Π of our ordinary angles.

The Speedy Photon in a Slow-Motion World

Suppose you pick up a nail using a magnet. Did you ever wonder how the force goes from magnet to nail? There is nothing that connects them! When you turn on a light, you instantly see light everywhere. How can something travel instantly? These strange events are called "action at a distance." One hundred years ago everyone was mystified that light, electric, gravitational and inertial forces appeared to take place instantly. That there was nothing between the interacting bodies made it even more confusing! Now we have an explanation of sorts: Forces happen between two bodies because "something" travels between them at the speed of light. The speed of travel of the "thing" has been well verified, but the "thing" itself is still not well understood.

We tend to feel a lot better about these quantum things if we create an imaginary picture of the quantum world. So scientists have decided to call the "thing" a photon. It is generated in an energy exchange process between the bodies, which conserves energy. We don't understand the *photon*, but the resulting calculations of the quantum world of electrons and atoms fit very well indeed with spectroscopic measurements and other evidence. So there is a lot of confidence in them. As a result we begin to feel certain that the real action of our world is taking place in the quantum realm. The human realm action we see is only that part which interacts with our five senses and our mind.

It has required about 50 years for scientists to realize that we live in a slow human realm where we never see all of what is happening. Behind the scenes, the real action of science is taking place in the quantum realm. It is like a movie which we think is a continual moving image, but is really a rapid sequence of tiny quantum photon events that happen very quickly, some of which we do not observe at all.

QUESTION FOR THOUGHT: *Think about whether there exists anything solid in the universe. Think about a solid metal and you realize it is not "solid," but composed of widely separated electrons and nuclei. Do we still have reason to believe there is anything at all which is solid?*

Big Things Depend upon Little Things. The main purpose of this chapter has been to illustrate the importance of the geometry of 3 dimensional space on the basic mechanisms of science and to suggest that if we want to understand more, space is one place to conduct our exploration. Especially important is the photon and the rules of quantum mechanics and relativity which govern the photon exchange between the particles of matter.

If you carefully reflect about the substance of the big things of our technological world: energy, mass, matter, force and power which make up our bodies, cities, food, machines, toys and bombs; you eventually come to realize that the tiny photon, whatever it is, lies near the heart of the whole business. The mysterious photon is a very simple means of exchanging energy between one place and another or between one particle of matter and another. How does it work?

Since we do want to understand more, in the chapters ahead we will continue to explore the nature of the particles and the photons which carry the energy between the particles and creates the forces which move our enormously larger human world. Our exploration so far appears to be leading us towards what at first appears to be *empty space*, and it tantalizing beckons us to search further among the properties of space itself. Perhaps space is not so empty after all.

The Photon by any Other Name is just as Confusing. Another purpose of this chapter has been to point out the importance of our human attitudes on the development of science. The *photon* is an excellent example. Before the photon had a name, action-at-a-distance was a subject of intense philosophical discussion and scientific enquiry. However, after the name was adopted most scientists heaved a great sigh of relief, and thought to themselves, "At last, we know how the action takes place!"

But think a bit: adding a name revealed nothing new at all about the "thing," and our scientific knowledge was not increased. However, we do feel a lot better about it and our curiosity is no longer so intense. For example, if a student today should ask, "What creates the force of electricity?" His teacher will reply, "photons." The student may be satisfied, but in fact, no explanation has been provided. There is something about having a label for things which makes most people feel that they understand, even when they don't.

"We all agree that your theory is crazy. The question which divides us is whether it is crazy enough."

– Niels Bohr

CHAPTER 6

Evaluating New Ideas

The Test of Prediction
Explanations are Good. Assumptions are Bad.
Arbitrary Constants are Often Only Band-aid Patches
Solving Enigmas Enhances Predictive Value
Many Birds With One Stone is a Good Shot
Simplicity is Nature's Way
How Probable is Probable?
Judging Your Own Ideas

CHAPTER 6
Evaluating New Ideas

Every new concept, experiment, calculation, or other piece of scientific work has to be tested. People, the scientific community, and the world all want to know if it is correct. There are accepted ways of testing new scientific work using logical and scientific viewpoints, and describing them is the main purpose of this chapter.

The best tests of truth should not rely on subjective factors such as style of writing, elegance of concept, notoriety of the author, or the prestige of the institution. Good work should be separated from bad using standardized, coldly objective methods. Some of the more popular testing methods of this kind are described in this chapter.

The Test of Prediction

If an experiment, a measurement or a theory is to be ultimately useful, it must have predictive capabilities. It should predict the results of yet another experiment or measurement, or it should determine the performance of new machines or apparatus. This predictive capability can be used to test the new work. To carry out a valid test, you have to predict something which is *not yet known by any other method*. Otherwise, you may be accused of false predicting; like dealing from a stacked deck of cards.

As an example, suppose you have a new theory of gravity. It enables you to predict that antiprotons have repulsive gravity and it provides you with a method of measuring the gravity. So you or someone else makes the measurement, and if it turns out repulsive as predicted, your theory is verified. Unfortunately, this one test is not very good, because there were only two possibilities, repulsive or attractive, and you had a 50-50 possibility of getting the right prediction by guessing. But suppose you predicted the + or - gravity measurement of six different particles, and they all turned out correct. Then the odds of getting the right predictions by luck would have been one part in 2^6, or 1 in 64. Most people would then believe your theory.

Another example, suppose you claim you can design a machine that will make 200 pretzels per minute. If you make the machine and test it, obtaining 216 pretzels per minute, you have an *absolute test* because there is no way of making pretzels by pure luck. Another almost absolute test would be to *predict the numerical results* of a measurement which hasn't been done before, and which can't be guessed or calculated. For example, this could be the lifetime of a neutron in a new strange environment, with an expected value different than the usual number (14 minutes). If you predicted, say 149 minutes, and the measurement was 146 (+or-4) minutes, then the probability of getting that result by luck would be much less than 1 part in 146. Your prediction is almost certainly verified.

Explanations are Good. Assumptions are Bad.

New theories are usually built upon *assumptions* which are required to create the mathematical or logical framework. Reason and mathematics tells us that as the number of assumptions needed to create the theory increases, so does the likelihood that the theory is invalid. On the positive side, if the theory provides *explanations* for phenomena which were previously unexplained, or provides new explanations to replace torturous difficult ones, this is valid evidence that the new theory may be correct. One way to make a judgement is to count the number of each of these opposing elements, *assumptions and explanations*, and then compute a "probability" that the new theory is correct.

A method of computing can be taken from the probabilities associated with flipping coins.

Suppose we count the number **p**, of explanations provided by the theory and compare it with the probability of flipping **p** consecutive heads, which is, Probability = $(1/2)^p$ x 100%. On the other side, we count the number n, of assumptions required, and compare them with the probability *against* flipping **n** consecutive heads, which is $\{1 -(1/2)^n\}$ x 100%. Putting these together, we can come up with an ad hoc formula:

Truth probability = $\{1-(1/2)^{p/n}\}$ x 100%   p > n

This is not the only formula one can choose, another possibility is:

Truth probability = $\{1-(1/2)^{p-n}\}$ x 100%   p > n

Unfortunately, explanations are often a personal judgement

CHAPTER 6/*Evaluating New Ideas*

Arbitrary Constants are Often only Band-aid Patches

New theories often contain numerical parameters that are needed to match experimental observations. For example, the gravitational force constant G is such a number, as is the electrical force constant. Both are constants of Nature. However, a zealous author may inject such numbers into his work and label them natural, when in fact he needs them merely to fix up his math so that it will match measurements. The more arbitrary constants are used, the less likely the theory is to be valid. However, constants are often necessary, so judgement is required.

The study of equations show that in order to completely define a parabola (a second degree curve), two constants are needed. A third-degree curve needs three constants, and so on. Thus a theory using a second degree equation would be valueless if it used two arbitrary constants because it could fit anything merely by adjusting the constants. But if only one constant is used, the theory has some predictive value.

Therefore, the total number of arbitrary constants should be counted, and used to reduce the Truth probability number above. In the same vein, an estimate of the degree of mathematical sophistication might be used to increase probability.

Less damaging to a new theory are *undetermined constants*: numbers which in principle can be measured in the future, but which for varying reasons are not yet available. In this case, final verification of the theory hangs in limbo awaiting the numbers.

Solving Enigmas Enhances Predictive Value

If a theory is propounded that provides an explanation for a previously unexplained rule of nature, or a puzzle such as the enigmas of this book, this is regarded as having predictive value. But since an unexplained phenomenon is usually known experimentally, the author of the theory knows ahead of time the value of the numbers he is explaining, so his explanation is not as good as a pure prediction of a new event, but still it is positive. You might choose to count them at half the value of a prediction.

The most potent explanations are those which come in a 2-for-1 package — a mechanistic description of the "inner works" of the phenomena, *plus* the mathematical apparatus necessary to predict. History has shown that the package rarely arrives full-flowered. At first, various persons gain an inkling of Nature's plan, then bit by bit the whole theoretical structure is put together in a complicated form. Finally, some complications are seen to be the result of accidently choosing a "hard way" to view the situation. Then, the clarity of hindsight allows us to create more simplified mechanisms and mathematics to replace the first clumsy ones.

It is often quite difficult to verify partial packages, yet that is the usual way science proceeds. It is easy to both attack and justify half-finished ideas. An antagonistic referee who has his own favorite concepts, will call a new idea for a mechanism, "half-baked" if it has no mathematical package. On the other side, its author, brimming with motherly love, will claim he has found the "heart of the matter." Similarly, new mathematical structures are often complicated and difficult to follow, so if there is no description of the inner wheels and works, it is hard to distinguish truth from mere lust to publish.

During the last 50 years, more and more of the frontier work in physics has provided only the mathematical half of the package. The reason is understandable — most of the work has taken place in the quantum or cosmos realms and our human realm intuition has been of little value in imagining the structure of the "inner works." As a consequence, many of this book's enigmas refer to the unknown mechanistic part of a package, for which a mathematical method is already known. Needless to say, this is frustrating! We feel like blind men who know the layout of the forest we walk in, but we cannot see the trees.

Many of the enigmas in this book involve *fundamental laws* and urgently need mechanisms to explain them. This fact places restrictions on the nature of new theories. If they are to be correct, they must agree with all the fundamental laws of nature which have already been found to be numerically correct. Agreement with them is a *necessary and sufficient condition* for correctness of a new theory. As has been pointed out, the number of basic laws are small, so the task of checking is not herculean. These fundamental laws for which we have no explanation, discussed in Chapter 4, are listed again below:

FUNDAMENTAL LAWS OF NATURE
1. Newton's second law, $F = dp/dt$.
2. Law of gravitation.
3. Conservation of Energy.
4. Law of electric charge.
5. Special relativity (constant light velocity, and any one Lorentz transformation).
6. Quantum mechanics (deBroglie wavelength, and the probability concept).
   And particles and their properties: electron, proton, neutron, neutrino, and photon, which obey the above laws.

One might quarrel that some exotic particles have been left out, but those do not seem to have a fundamental role. To the contrary, it appears they are only excited states of the ones listed above, to which they will eventually decay. One can also question whether the photon even belongs here since it is derivable from the laws 4, 5 and 6, and its designation as an entity is debatable. The same may be true of the neutrino.

Because these laws are *fundamental*, it follows that nothing more basic is known, from which they might be derived. All of them are also on the *enigma* list because they lack a description of a mechanism or "inner works."

Many Birds with one Stone is a Good Shot

A new theory may gain plus points in its favor by explaining other enigmatical behavior in addition to its primary target. For example, a theory to explain how cookies disappeared from the kitchen might be significantly buttressed if it also explained the loss of apples and peaches from the backyard trees.

Simplicity is Nature's Way

In the past, progress in science has been helped by assuming the *simplicity postulate*. This is the point of view, adopted in view of our ignorance of the future and purely as a working hypothesis, that the structure of the universe is as simple as possible. Thus, if we have two competing explanations of some measurement, for which the arguments are equal, but which involve equations of third and second order respectively, then the one of only second order is probably more correct.

A historical example is the case of the Copernican theory (Sun at the Center) and the Aristotelean theory (Earth at the center) competing to explain movement of the planets. Both theories could assemble the required mathematics needed to predict the measurements of planetary positions. However, the complexity of the seven Aristotelean "epicycle" equations was appalling, and the number of *arbitrary constants* was even larger. When Newton was able to show that only *two* equations $F = ma$ and $F = Gm_1m_2/R^2$, and only one arbitrary constant G, were needed for the Copernican theory, the judgement was overwhelming in his favor. Furthermore, his theory also explained other paradoxical phenomena, such as the motion of falling objects and the path of a projectile.

How Probable is Probable?

Nearly everything in life including scientific calculation is a gamble and has a chance of going wrong as well as right. Science is not infallible. For example, there is a chance that all the air molecules moving randomly in your room will all end up in the corner during the next five minutes, and you will have nothing to breathe! This chance is astronomically small but it is possible. By chance we mean something like a guess, except that when we deal with science we try to learn as much as we can about the accuracy of our guesses.

The mathematics of probability deals with these chances but is too big a subject to discuss here, so we will point out a few rules of thumb which provide you with a start in this part of the game.

DEFINITION: Probabilities always add up to 100% or 1.00.

For example, when flipping a coin, the probabilities are :
heads   0.50 probability
tails   0.50 probability
Total   1.00

FACT: Probabilities can be multiplied to find the final probability.

If you want to know the probability of two heads in a row, multiply 0.5 x 0.5 = .25. In the long run you will get two heads 25% of the time. Eight heads in a row will probably happen $(0.5)^8$ = .039 % of the time, and so on.

Many common probability calculations can be roughly estimated using an interesting situation called *The Random Walk*. I will tell you about this trick and leave the rest of probability for you to find in a book, when you need it. The random walk goes as follows:

Suppose a jumping frog is placed on a rock in the middle of a large empty field and he doesn't know which way to go. So he just hops randomly to see what happens. Can you predict how far from the rock he will be at any given time? Probability provides the simple answer that if he makes N hops, then the probable distance from the rock will be the square root of N hops, as in Figure 6-1.

You can use this neat rule to estimate many situations like: How fast does perfume travel through the air to its target noses? If you know the time to get to the first nose, the above rule will help you estimate the time to the rest of the noses. Another use for this rule is to estimate the probable error of statistical samples. For example, suppose you flipped one coin a 100 times, and you want to know how close you will come to 50-50 heads and tails. The square root of N rule will tell you. Plot a dozen sets of 100 coin flips as in Figure 6-2. The width of the peaked curve is close to $(100)^{1/2}$ = 10. It is called the *statistical root-mean-square deviation*. If you want the + or - value of the deviation, it is *half* the width.

Statistical deviation = $1/2 \sqrt{N}$

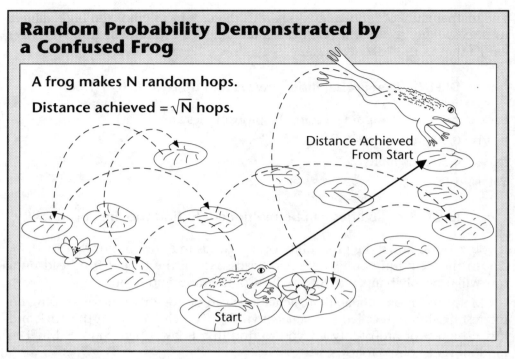

Figure 6-1. The magic of $\sqrt{N}$
When calculating probabilities connected with N events or measurements all of the same kind, the $\sqrt{N}$ is often the most useful factor. You can use it to quickly estimate probabilities in your head. In this figure it tells you how far the frog has probably traveled from his starting point.

Other examples abound, such as:

1) If 1000 sandwich-eaters are sampled and 401 like rye bread, what is the error (deviation) of the 401 result? Answer: Probable error (no. of rye eaters) = $\pm\, 1/2\, \sqrt{N}$ = $\pm\, 1/2\, (401)^{1/2}$ = $\pm\, 10$.

2) Suppose you wanted to be very careful with a length measurement, so you did it 64 times. Then you calculated an average. Then you found the deviation from the average of each measurement. What is the error of your average measurement? The error is proportional to $1/(N)^{1/2}$, that is

Measurement Error = average deviation/$\sqrt{N}$

In this example, $N = 64$ and $\sqrt{N} = 8$.

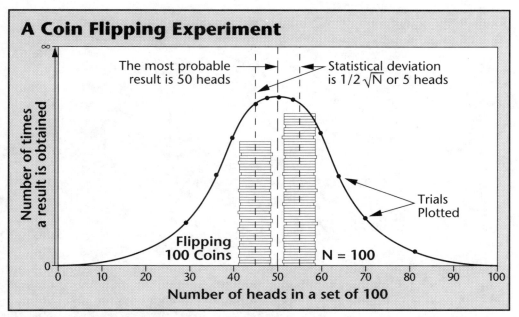

Figure 6-2. It is easy to demonstrate statistics
You can quickly show yourself how the $\sqrt{N}$ Statistical Law works. Get 100 coins. Flip them about ten times, counting the head and tails. Then plot this figure to see how your experiment matches.

Exploration in science is often a lonely road, and you will find few people willing to listen to your new idea, much less evaluate it for you. You are lucky if you have such a friend. Most other scientists are busy working on their own favorite topics and don't have time to look at yours. It might even happen that your ideas compete with theirs, and then you may find that your ideas are not welcome. In such a case, it becomes important to be able to judge your own ideas and decide yourself if they are right or wrong. After all, you don't want to keep working on a concept which is not going to bloom. It is better to find out as soon as you can, and start work on another more fruitful endeavor.

It can also happen that your good idea has an obvious possibility of becoming a money-earning commercial success or, in the academic world, good enough to win a lucrative research grant. In this case, you are well advised to make an early judgement of it, so that you obtain some reward, for which you probably have worked very hard.

Judging your own ideas

"The most incomprehensible thing about the universe is that it is comprehensible."

– Einstein

CHAPTER 7

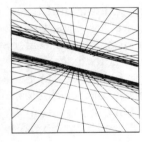

Space, The New Frontier

The Properties of Space
Space in the Quantum Realm
Curved Space is Well Known in the Cosmos Realm
Geometry is not Mathematics, but a Property of Nature
Practical Navigation was the Precursor to Non-Euclidean Geometry
Four Dimensions in our Universe?
The Space of Einstein's General Relativity is Distorted
    by the Mass Within it.

CHAPTER 7
Space, the New Frontier

The title of this chapter is a guess that the next leap forward in basic physics will be a result of the study of space — that stuff in which we all exist but know little about. We are constantly finding our inquiries come to a baffling halt at the portals of space, so it follows that new answers may come from more knowledge of space.

The Properties of Space

Nearly all the theories of matter and the universe implicitly assume that "empty space" has distinct properties. However, few searches have been made to directly investigate those properties. If you seek information about space, you will find bits here and there embedded in the axioms and basic assumptions of other theories. Some mathematical works are concerned with space's geometric properties. Almost no experimental data exists. Yet, the search for explanations of fundamental phenomena constantly points to space, so we suspect that the fundamental laws and the constants h, c, m,e, and G attached to them are somehow determined by space. Hoping that this suspicion will lead somewhere, let's examine what little is known of space. This information is typically involved with either the Quantum Realm or the Cosmos Realm, so I will divide it into these two categories.

Space in the Quantum Realm

In the quantum realm, we have almost no experimental knowledge of the geometry of space. The field has attracted little interest, so that for dimensions less than about 10^{-10} meters, our knowledge is almost nil.

In "solid" matter the atoms are not attached to one another. Instead the atoms are held in their rigid lattice only by forces in the space which separates them. We can easily calculate the forces between the atoms of crystalline materials. By human realm standards they are enormous. There seems no choice but to endow space with a property of great rigidity. If this idea is extended to the nuclear domain, where forces are a thousand times larger, the apparent rigidity of space between nucleons exceeds anything we can imagine. Space appears to be very strong.

It is surely an important question to ask if space density can vary on a quantum scale. The first serious suggestion that space was non-uniform on a small scale was by the English geometer, William K. Clifford (1845-1879), who wrote,

> "I wish here to indicate a manner in which these speculations may be applied to the investigation of physical phenomena. I hold in fact:
>
> 1) that small portions of space are in fact of a nature analogous to little hills on a surface which is on the average flat; namely that the ordinary laws of geometry are not valid in them.
>
> 2) that this property of being curved or distorted is continually being passed from one portion to another after the manner of a wave.
>
> 3) that this variation of the curvature of space is what really happens in that phenomenon which we call the *motion of matter.*
>
> 4) That in the physical world nothing takes place but this variation, subject (possibly) to the law of continuity."

His intuitive sense was a remarkable precursor of quantum mechanics, still 50 years away! He attempted to use these ideas to explain optical double refraction.

Clifford was a remarkable mathematician. In a tragically short life of only thirty-four years he enlarged scientific thought by a series of contributions as original as they were fertile and as far-reaching as they were lucid. He taught that applied geometry is an *experimental* science, a proper part of physics. In his time this was a radical view, even today this concept is not yet fully absorbed.

QM Theory already uses Space Properties. The theory of quantum mechanics is not predicated upon any property of space at all, but space ideas are frequently used to "justify" mathematical procedures, and these justifications imply a surprising amount of detailed space properties, including:

> 1) Quantum waves can travel in space. This implies that space is a medium.
>
> 2) Particle pairs can be created out of space which exist fleetingly, for a time $t = h/\Delta E$,
>
> where ΔE is the pair mass-energy. *This implies that the structure of particles is only space.*
>
> 3) Virtual charge can be induced in a small region of space around a real charge, which may partially neutralize the charge. *This suggests that charge is a property of space, rather than particles.*

These ideas have been used as illustrations for QM rather than facts, and no experiments have been made to see if space does have such properties. The justification by QM theorists is that the resulting mathematical approximations give the right answers. But if you choose to believe them, you must also conclude that space has real properties, of which we understand only a little.

Can space be stretched or sheared? Suppose it is true that space can be distorted to form a structure of some sort, maybe a particle. Does this demand certain additional properties of space? Topologists, the geometers of space, have frequently suggested that it does and one required property is *continuity* of the space. This means that any structure or motion of the particles must not destroy the continuity of coordinate lines which map the space.

For example, suppose three sets of mutually perpendicular grid lines are used to locate position in an x, y, z cartesian coordinate space. Then if the space is twisted in some way to form particles, the twisting must not break any of the coordinate lines. Of course, it can be assumed that the space and lines are almost infinitely stretchable. The continuity rule also says you may never separate the particle from the space which forms it.

An example of allowable motion of space is waves which undulate by moving the space back and forth, always returning to a previous position. Such motion never moves perpetually in one direction, which would distort the space to an extreme. Thus the coordinate lines of the space always remain continuous. We conclude that ordinary types of waves allow continuity of the space in which they propagate.

An example of an unallowed motion is a particle with a spin like a wheel on an axle. Space would get all wound up, being stretched at the axle. The coordinate lines used to map the whole space would become twisted and stretched without limit as time passes. Eventually, the space will become "ripped," which is not allowed because then the lines would not be continuous. You might suggest that the space could be constructed of two sections which slide past each other at a plane of discontinuity, but this is also not allowed because there must be only one region for all of space.

The Rotational Space Group SU(2) describes the Dirac Equation. The SU(2) group in Chapter 5 which described *spherical rotation*, displays many properties which link it to the quantum realm. Spherical rotation is different from cylindrical. It takes place around a point rather than an axis. The rotary motion, if we apply it to space rather than physical objects, does not result in the accumulation of energy since only space is rotating, not matter as we perceive it.

The motion preserves the continuity of space. If you tie a rope to the rotating body, the rope will *not* wind itself up into knots! On the other hand if you tie a rope to an ordinary (cylindrically) rotating body, the body must stop rotating or break the rope. But this destruction of the continuity of space is *not* found in spherical rotation.

What does this rotation property imply? It suggests that if a particle were constructed of the fabric of space, then spherical rotation makes it possible for the particle region to rotate and maintain continuity with the rest of space.

What happens to the space around the spherical rotating particle? If you carefully watch your rotating cork model which you built in Chapter 5, you will see that as the rubber bands are stretched, the space is stretched and lines of continuity oscillate in a wave-like fashion around the center. But the frequency of these strain-waves in space is just *half of the frequency* of the rotating object, as we saw in Chapter 5. But this factor of *one-half* is also found in the quantum realm, because in solutions of the Dirac Equation which describes an electron in Chapter 9, such a factor appears when calculating the "spin" of the electron. In fact, we know there is a direct relationship, even though spin is hidden in the complex math, because the SU(2) space has been shown by E. Battey-Pratt and T. Racey in 1980 to describe the space of the Dirac Equation and its solutions. Since the Dirac Equation is the best available description of an electron, it provides us with more hope that this is the right direction in which to search.

Curved Space is Well-Known in the Cosmos Realm

General relativity presumes that space can be curved due to the presence of matter, resulting in distortion of Euclidean geometry. If space is strongly curved by concentrated mass, the result is the famous black holes. Space is thought to be slightly curved on a large scale throughout the universe, so that light rays can curve back to their origins. A tiny curvature of light rays passing the sun has been measured.

It is surprising that so little attention has been given to investigating space, in view of how often it is called upon to bolster other ideas. But this is changing in the age of space vehicles. Research projects in the cosmos realm have begun with telescopes and measuring instruments on satellites. This has lead to initial estimates of the size and density of the universe, which are important to general relativity, a theory which requires measurements of mass distribution in the universe to provide a meaningful numerical basis.

If we estimate the energy density of the universe as we roughly know it, we find surprisingly that the total energy of gravity may be exactly equal to the total energy of all mass particles. This surprising result may have deep meaning. Recent measurements of more distant stars deep in space, portend the beginning of significant knowledge. At these distances there are strange objects: pulsars, quasars, blazers, and gravity lenses. Are these objects clues to final understanding of space, or are they just new objects in the celestial zoo?

Although experimental physics in the past has had little involvement with space, natural philosophers have always been intrigued by space and cosmology over the centuries. Their recent opinion has changed it from a branch of pure mathematics to a property of the physical world. Philosophy and hard science have joined hands on the path of exploration.

Geometry is not Mathematics, but a Property of Nature

Euclid, in his famous treatise on plane geometry, Pythagorous and other Greek and Arabic mathematicians who laid out the foundations of modern mathematics, all made apparently unconscious assumptions that the structure of space was homogeneous or 'flat.' Such an assumption is now given the name "Euclidean.' They believed that geometry was purely a mental art, and together with mathematics, could be completely understood by abstract thinking. To them, geometry was a product of the human mind and had no relation to the physical world.

Three thousand years passed before any other view of geometry was imagined. In the nineteenth century, the assumptions of Euclid were examined carefully, challenged, and alternatives discovered. The first glimmer of a connection with physical science had begun.

One of the first to recognize that geometry was a physical property of Nature and not an art of mental abstraction was William K. Clifford (1845-1879). In numerous lectures and papers he pointed out that Euclidean geometry is based upon measurements in a very restricted space — our laboratories on Earth. He argued that our knowledge of the physical world *and* its geometry must be based upon actual measurements. He discussed the assumption that the angles of a triangle add up to 180°, and pointed out that we cannot logically accept it in outer space, because known measurement accuracy of the stars does not justify it. Clifford summarized our very limited measuring capability in outer space when he wrote:

> "Now suppose that three points are taken in space, distant from one another as far as the Sun is from alpha-centauri, and that the shortest distance between these points are drawn so as to form a triangle. And suppose that the angles of this triangle to be very accurately measured and added together ... Then I do not know that this sum would differ at all from two right angles; but I also do not know that the difference would be less than ten degrees, or the ninth part of a right angle."

That is, he pointed out that astronomical 3D measurement error is more than 10°, so we have no evidence to say that outer space is the same as our space.

His contention is still valid since astronomical accuracy at large distances has not improved significantly. We do not know today how much distortion of

Euclidean geometry may exist in outer space. Remarkably, his prophetic papers were published forty years before Einstein announced his theory of gravitation (General relativity).

Other geometers of the time agreed with Clifford, and paper and pencil research of space began. The success of their challenge to established ideas of the nineteenth century geometers, has culminated in the cosmos-realm concepts of energy, time, and space, summarized by Einstein in his two relativity theories. Yet, these new concepts are still unsupported by measurements.

Practical Navigation was the Precursor to Non-Euclidean Geometry

The problems of navigation on the spherical surface of the Earth became important to commerce in the seventeenth century, and governments brought this to the attention of mathematicians.

The problem arises from the Euclidean assumption that the ratio circumference/diameter of a circle is always exactly π. As we all know, this not at all true in the spherical geometry of the Earth's surface, and a practical theory for navigation was urgently needed as nations began to explore the seas. Geometers were employed.

After developing the basis of spherical geometry to earn their salaries, the geometers extended the work to multi-dimensional spaces greater than three, apparently as an academic exercise. In these higher spaces the assumption that the ratio of the volume of a sphere to its radius cubed equals $4/3\,\pi$, is *not* necessarily true.

The first to recognize that geometry was an intrinsic property of a surface was Karl Friedrich Gauss (1777-1855). He discovered that the curvature of a surface can be completely determined solely from the distortions present in a flat chart of the surface; like ascertaining the shape of the Earth from a map of the Asian Continent in an atlas. Gauss was one of the greatest of all mathematicians, contributing an enormous amount and variety of new mathematics to the world. He and other nineteenth century geometers, Lobachevsky, Clifford, and Reimann, developed the mathematics of non-Euclidean geometries. Seventy-five years later Einstein made extensive use of their work to frame his General Relativity.

Four Dimensions in our Universe?

There is scant physical evidence that geometries higher than three exist in this universe. Nevertheless this is a popular idea. On the supposition that measurements might show deviations from the formula, volume = $4/3\pi R^3$, some science-fiction writers describe the existence of worlds of four dimensions, in analogy to the tale by Abbot in his book "Flatland," where two-dimensional creatures lived on the surface of a sphere in three dimensions. Another reason for the popularity of four dimensions is the ease with which a mathematical

CHAPTER 7/*Space, The New Frontier*

description can be generated. You simply add a fourth variable and treat it like x, y, and z. For example, one can write the length of a line, **L**, in any number of dimensions, as follows:

2D space	3D Space	4D Space Etc.
$L^2 = X^2 + Y^2$	$L^2 = X^2 + Y^2 + Z^2$	$L^2 = X^2 + Y^2 + Z^2 + W^2$

On the other hand, experimental evidence does exist to suggest that space is not homogeneous. But, setting aside the needs of science-fiction writers, there is no reason to imagine that we live unknown to ourselves in four dimensions, or any number of dimensions except three. Mathematicians are quite able to describe the properties of a non-uniform space and at the same time keep themselves and us living comfortably in three-dimensional arm chairs.

The false logic of the people who make an analogy, à la Flatland, between three and four dimensional worlds can be easily seen. The analogy suggests that the space of a system of galaxies has no boundaries because it resembles the surface of the Earth, but only one dimension larger. They reason that the Earth's surface has no boundary because it curves around into a third dimension. We are then tempted to quickly infer that a three-dimensional space must somehow curve around into a fourth dimension. This inference is unwarranted because all the measurements required can be carried out entirely within 3D space itself. It is *not*, for example, like a navigator on the Earth who uses the position of the Moon, which is outside of the surface, to find his position.

To further emphasize lack of a need for a fourth dimension, consider the information provided by an atlas of the world. It is a book full of flat maps with longitude and latitude lines marked upon them. For the surface traveler, the maps provide complete information about the spatial relations of everything on the surface. This conclusion was an important result of the work of Karl Friedrich Gauss. There is no need for aerial views. Similarly, Einstein's theory plus a set of measurements of matter in space provide a complete atlas of cosmic space, without a fourth dimension.

Non-homogeneous Space is at Center Stage. The little amount of experimental information now available is easily interpreted using a non-homogeneous space, rather than four dimensional space. This concept has practical utility.

The reasons for considering a space that is non-homogeneous relate to problems in both the quantum and cosmos realms. In the quantum realm, the concept of point particles unhappily leads to fundamental contradictions about energy. Therefore, ideas have been sought, using non-uniformity of space, to find a structure for "particles" without using a point mass or charge. In the cosmos realm, human minds are unable to grasp a concept of "infinity".

We cannot understand questions like, "What happens if you travel forever? Where does an infinite journey end?" But mathematicians have found answers to these questions, if they use a non-homogeneous space in general relativity.

The Problem of Cosmological Infinity. Is the universe finite or infinite? The notion of infinity has always been wrapped in mystery and triggers our apprehensions. Just to understand infinity, leaving aside whether it exists or not, is an achievement.

The Celestial Bodies Embedded in Crystal Spheres

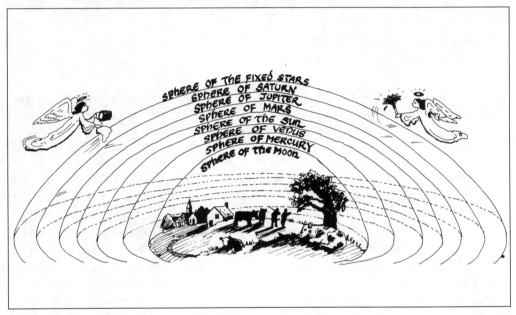

Figure 7-1. The spheres of Heaven
This concept of the structure of the Universe was held by the Catholic Church and most learned persons for about 1500 years. Cleaning the crystal spheres was a regular duty of the angels. Luckily, no one asked why falling meteors didn't shatter the crystal!

Aristotle's picture of the universe, shown in Figure 7-1, was widely accepted in medieval Europe. In spite of its popularity, it was open to one devastating objection: What is outside of the outer boundary? The question of boundedness is vital to comprehension of infinity in cosmology.

After Aristotle, during the European Age of Reason, Euclidean geometry was the main instrument in attempting to understand infinite physical space. Leibnitz and Newton shared the view that space was both infinite and

Euclidean. They disagreed on how matter was situated in space — remember, they had no significant telescopic measurements. For Leibnitz, a group of stars was unthinkable; such a group would have to be in some specific location in space and God would have had no sufficient reason to put it in one place rather than another. Leibnitz thus concluded that the universe must be infinite. Newton rejected that possibility on the grounds that God is the only possible actual infinity. Although today these arguments do not seem persuasive, at that time they were considered sufficient.

A century later, observations using large telescopes provided the first view of our galaxy plus a few other galaxies and objects. At the same time, the non-Euclidean geometries provided a conceptual framework for newer ideas. Building upon these, Einstein's General Theory of Relativity created a new geometry of space and a new approach to the question of infinity.

The Space of Einstein's General Relativity is Distorted by the Mass Within it

His theory deals with the properties of matter and its influence on the geometry of space. It regards a galaxy as a typical unit of matter on a cosmic scale with all the galaxies affecting space in an averaged fashion. The structure of space, that is, the shape of the coordinate grid lines, is influenced by the presence of matter, so that one can say that space takes on a "curvature" dependent on the distribution of matter throughout the universe. It is the curvature of the coordinate grids by matter which makes a finite universe possible in the theory. Measurements of the amount and distribution of real matter must be injected into the theory in order to draw conclusions about reality. Such measurement have only barely begun, so it is still a guessing game.

The firm experimental reasons for believing that space is non-homogeneous on a large scale are two: 1) the measurement that light rays curve if they pass close to the Sun (space is distorted by a massive body), and 2) the slight precession of the orbit of Mercury (The space grid lines of the orbit are distorted by the Sun.) Both of these measurements were inspired by the theory of General Relativity.

The connection between curvature of cosmic space and a finite world can be explained by an analogy with what happens on the curved surface of the spherical Earth. Imagine that the Earth is completely covered with houses. The earth surface corresponds to cosmic space and the houses correspond to stars in space. Observe that every house has a neighbor and you can travel forever from one to the next without end, because the earth is round. At the same time, the total number of houses is a definite number. This surface is thus *unbounded* but nevertheless finite. Note that finiteness and boundedness is still true even if the Earth were distorted with bulges and depressions, or stretched out in various ways. It is also true for the analogous cosmic space, whose structure is still totally unknown to us.

This analogy between three-dimensional space and the Earth's surface is quite close. To change from the surface to space is merely a matter of suitably increasing the number of mathematical variables (2 to 3). But to see a "picture" of the situation requires a vivid imagination. Don't feel you are alone if you can't see the picture. Some can, some can't.

We have insufficient measurements or other information to tell us whether the universe corresponds to a neat tidy ball, or a Christmas cake full of fruits and nuts, or the weird twisted creation of a surrealist artist.

Einstein's theory only provides a method and the mathematics to deduce the structure of cosmic space, only *after* measuring the matter in it. The theory can't tell us yet how Nature really did it.

Facts are no match for belief.

— witticism

CHAPTER 8

All About Waves

The ABCs of Waves
Waves are Produced by Energy Exchanges
A Typical Wave is Sinusoidal
Vector Waves in 3D Space can be Polarized
Intensity of Wave is Proportional to the Square of the Amplitude
Electromagnetic Waves are Special
You Never Know a Wave Until You See it, Hear it or Find it!
Waves Travel Peacefully when the Media are Linear
The Magic of Complex Numbers can be Applied
    to Sinusoidal Waves

CHAPTER 8
All About Waves

Waves surround us in our daily life: They are the music and other sounds we hear, the light which conveys the images we see, the heat flow which makes us cold or hot, waves in our nerves and brain which convey our emotions, and a myriad of other sensory events. If we closely and microscopically examine all the matter and phenomena that we can, we discover that waves of some sort are always involved. Are waves a basic part of the universe? We have no proof, either way. If we must choose a single assumption to explain the forces and laws of Nature, a good guess is that waves are the means of communication of all forces, and that waves are probably the basis of the laws.

How good is that guess? The answer is that our knowledge, and our certainty, depends on the type of force or law. In the case of electrical forces such as those of the electrons in the atoms, we are very certain, because our eyes and skin can detect the waves of the energy transfers as do our laboratory instruments. We are less certain of other forces such as gravity, because we have no instruments which can detect the possible waves. We don't know the means of gravity-force communication. We can only speculate that it is a wave.

In the case of special relativity, the basis of this fundamental law involves the properties of light, which is an electromagnetic wave. Similarly, the fundamental laws of quantum mechanics are wave laws, although we are not certain about this kind of wave because, again, it is outside our range of detectability.

In any case, for the adventurer who is exploring the universe, the matter in it, and the laws which govern it, a knowledge of waves is important.

The ABCs of Waves

Before saying anything more about waves, we need a vocabulary to describe them. Referring to Figure 8-1, you should learn the following technical words, and their common symbols:

WAVELENGTH (λ)	=	The observed distance between two adjacent wave crests.
WAVE VELOCITY (v, c)	=	The speed at which crests pass by an observer.
FREQUENCY (f)	=	The number of crests which pass an observer each second.
AMPLITUDE (A)	=	The size of the wave measured from zero to the peak.
PERIOD (T)	=	The time for one wavelength to pass an observer, T = 1/f.
POLARIZED WAVE	=	A wave with amplitude in only one direction, say x or y.
MEDIUM	=	The material or substance in which a wave travels.
RESONANCE	=	Energy exchange between two oscillators of the same frequency.
STANDING WAVE	=	A stationary vibration. A combination of two moving waves.

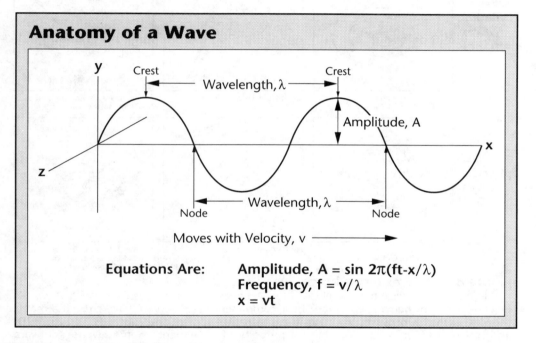

Figure 8-1. Learning to talk about waves
    This typical sinusoidal wave is labeled with the accepted terminology used to describe waves. How do you know this wave is polarized in the y-direction?

An assortment of waves found in the natural world around you are shown in Figure 8-2.

CHAPTER 8/*All About Waves*   107

Waves That Occur in Ordinary Objects

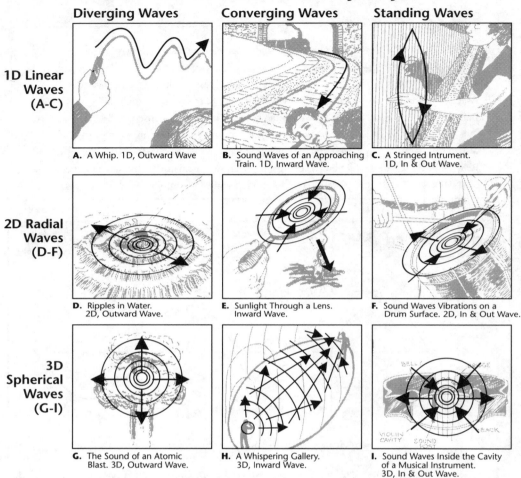

Figure 8-2. Waves move in various ways.

In order to visualize the complexity of wave types, it helps to classify waves that we can perceive with our human senses into their component parts. So we have created a matrix showing examples of waves vibrating in each of three dimensions: linear (1D), radial (2D) and spherical (3D). Waves also have a property of movement. Some diverge (outward moving) from a point and others converge (inward moving) and some are standing which are actually two waves moving in opposite directions (in & out).

It is easy to find examples in nature of 1D, 2D, and 3D standing or diverging waves. But the reader should note that waves only converging to a point do not exist in nature, so it is especially hard to visualize. Instead, we have included three examples of waves that have some diverging properties. This should come in handy when visualizing the Space Resonance Theory which we will introduce in Chapter 12.

From these definitions and figures you can now deduce the simple rule that velocity equals wavelength times frequency. That is:

$$\lambda f = c$$

Different kinds of waves travel at different velocities. Sound waves travel in air at 1100 feet per second, and electromagnetic waves travel in space at c = 300,000,000 meters/second.

Waves are Produced by Energy Exchanges

If you are curious, you might ask, why do waves occur? And, if they are universal, what universal situation leads to their production? The answers to these questions require us to study the medium in which the waves travel. Waves do not exist without a *medium* which determines their structure. For example, a stringed musical instrument wave needs a stretched string as its medium. Sound in air or waves in musical pipes need air under pressure as a medium. Waves on the surface of a lake travel in water which is pulled downward by gravity.

The media always have two properties: 1) Motion of the media creates kinetic energy of motion, and 2) Displacement of the media creates potential energy of displacement. Many types of media exist and many kinds of waves exist, but all have these two properties.

As the wave proceeds in its travel, there is a continual exchange of potential and kinetic energy in time with the beating of the wave. Examples of the energy exchange are easy to find. Consider an underwater elastic membrane enclosing an air-bubble connected to an air pump. This is an underwater sound source. The pump alternately increases and decreases the air in the bubble. When the bubble is changing its size, the water surrounding it is in motion and thus has kinetic energy. When the bubble enlarges, the compressed air inside has potential energy. The oscillating bubble continually exchanges its energy between these two forms. It will generate a water wave that travels radially out through the water.

The wave traveling along a string has kinetic energy (KE) due to the moving mass of the string and potential energy (PE) because of the stretch of the string. An electromagnetic wave exchanges its energy between electric (PE) and magnetic (KE) fields. However, this example is hard to visualize.

When a wave begins at its source, energy is abstracted from the source due to the oscillating behavior of the source. And when the wave reaches another oscillator, this energy can be delivered to the oscillator, but only if there exists a mechanism to enable the waves to transfer energy to it. An exchange mechanism is essential, and its concept is of fundamental importance.

Space is the common denominator of Waves. It is curious that despite the great variety of wave types, such as mechanical waves in metal, light waves in space, sound waves in air, water waves, etc.; if you analyze them to find what is happening on a microscopic level, you discover that they all appear to be related to the properties of space. This is because the medium of all mechanical waves is matter composed of atoms. The KE of the waves is actually KE of the relative motion of the atoms, and the PE is from the forces between them. This path of analysis leads to the same puzzle, the role of space, that we encountered in Chapter 2 where we sought to understand the meaning of length, time, and mass.

The puzzling relation between waves and space can be seen as follows: Measuring wave displacement to determine displacement (PE) energy demands that we understand the meaning of length. When we attempted this in Chapter 2, we were forced to conclude that length is a property of space that we do not understand. Similarly, if we want to determine motion or velocity to find kinetic energy, we must understand the rate of change of length with time, and in Chapter 2 we had to conclude that this also had something to do with space in a puzzling way. Finally, in the case of electromagnetic (e-m) waves, to which other waves can be reduced, scientists have found by experimental measurement that energy and frequency of the e-m waves are interchangable by using appropriate constants. Electromagnetic waves travel in space.

A Typical Wave is Sinusoidal

It is commonly observed that waves of pure tone or constant frequency have the form of a mathematical sine or cosine as in Figure 8-1. How does this occur? It is because the forces in the oscillating media are balanced at all times. If forces are balanced, the following must be true:

Rate of change of KE with respect to velocity =
rate of change of PE with respect to displacement.

When you write this rule using the calculus of Chapter 2 you get a *wave equation* whose solution is always a sine or cosine function, such as

$$\text{Amplitude} = A \sin 2\pi(ft - x/\lambda)$$

The **A** in the equation represents the maximum amplitude of the wave. The frequency of oscillation is **f**, according to time **t**, and λ is the wavelength. The (-) sign in the parentheses means the wave is traveling along the x axis in the positive direction. If a (+) sign is used, the wave travels in the opposite direction. We conclude that the sinusoidal property is equivalent to the property of exchanging KE and PE energy back and forth.

Vector Waves in 3D Space can be Polarized

If the wave amplitude also has a direction, then the 3D character of our world has entered the picture. We describe this by *polarization*. For example, the wave of an electric field represents the amplitude and direction in which a charge would be forced. If the electric field points in one of the 3D directions, either x, y, or z, that direction names the polarization. If a wave is moving in, say the z direction, and the field is pointing in the y direction, one says the wave is *y-polarized*.

In Nature, nearly all electric waves are polarized in directions perpendicular to the direction of travel. If the wave travels in the z direction, the wave could be either x-polarized or y-polarized. Ordinary light waves from the Sun, fire, or candles are not polarized. Instead they are random mixtures of both polarizations which on the average have no net polarization. Light reflected from a smooth surface is always polarized because the surface suppresses the perpendicular component. Polarized eyeglasses can reduce the glare from such surfaces.

An arrangement of two polarized waves traveling in the same direction can result in a total wave vector which twists like a helix, called *circular polarization*. The wave vector rotates counter-clockwise like a left-handed (L-H) screw if the two waves are perpendicular to each other and the phase of one of them lags behind the other by -90°. Similarly, if the phase difference is +90°, the rotation is clockwise like a right-handed (R-H) screw. This property is often used in satellite microwave radio to avoid interference of signals since a receiver that receives L-H circular waves cannot receive R-H circular waves.

Scalar Waves have no Direction. Some types of waves have no direction. The amplitude of the wave represents just a number called a *scalar*. For example, sound is a wave of varying air pressure. Along the path, the pressure rises and falls above the ambient pressure of the normal air, and the amplitude is the varying value of the pressure, a number varying sinusoidally. Quantum waves are of this type too.

QUESTION FOR THOUGHT: *What about a wave on a string? Is it a vector or a scalar wave? Is it polarized?*

Reflection of a Wave occurs When the Media changes Properties. When a traveling wave encounters a change of the properties of the medium in which it travels, all or part of the wave is reflected. The reflection of a wave can be geometrically represented using a superposition of two waves, traveling in opposite directions. Frequently, like the image behind a mirror, one of the waves is not really there. But there may actually be two waves.

Look at the simple example of a *one-dimensional* wave on a string. Suppose that one end of the string is held by fastening it to a solid wall. This results in the condition that the displacement of the string must be zero at the wall. If the wave is traveling towards the wall, then a mathematical way of keeping the amplitude at zero is to assume the existence of a second wave traveling away from the wall whose amplitude is equal and opposite to the first wave at the position of the wall. It now would make no difference if we clamped the string at just that one point to represent the wall. The two oppositely traveling waves must be sinusoidal and must have the *same wavelength* in order to satisfy the condition of zero amplitude at the wall.

A Standing Wave is a Combination of Two Waves. Consider a different situation. What happens if the string, of length L, is clamped at both ends? Then the amplitude of the combined waves must be zero at *both* ends. This can only be true if sin(2πL/λ) = 0, or L = nλ/2 where n is an integer; which means the string length is exactly half the wavelength. Musically, this says that if n = 1 the string will vibrate at the natural frequency of the string, and if n = 2 the frequency is double, an octave higher. If n = 3, it is another octave higher yet, etc. Thus the string can vibrate at many possible frequencies, provided the string length is an integer multiple of the string size. Also more than one frequency can occur on the same string.

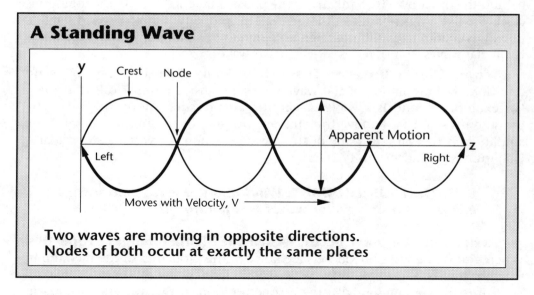

Figure 8-3. Two moving waves make one standing wave.
If two oppositely-moving waves of the same frequency pass each other, the medium appears to move to-and-fro without longitudinal motion. A violin string is an example. The two waves are also reflected back and forth at both ends of the string by the bridge and the nut. The length of the string must be an integral number of half-wavelengths.

Sometimes there are two oppositely-traveling waves on the string. Their amplitudes will combine, and if the wavelengths are the same, the appearance of the string is a to-and-fro vibration, with no apparent traveling motion. This is called a s*tanding wave* as in Figure 8-3.

Waves can also travel simultaneously in *two dimensions*; for example across the surface of a drum-head. Standing waves can occur, also in two directions across the drum-head. A condition imposed upon the waves is that amplitude must be zero all along the edges of the drum. This leads to mathematical complexity for the shape of the wave patterns but the zero amplitude idea is the same as the string. Unlike the string, the wave sets are 2 dimensional and each set may have harmonic components.

Three dimensional standing waves can occur in volumes of various shapes, for example, within the body of a violin. Because of the complicated shape of the violin, an analysis by mathematics would be too complicated to be useful even if it could be solved, but you can be sure that the best violin makers know where the standing waves occur and what their relative amplitudes are. They have learned by experience using their eyes, ears, sense of touch, and brain, but without math!

Reflections in an open pipe. Clamping a string or drum-head is not the only way that the properties of the media can be changed. When a wave travels down a pipe such as a trombone, a reflection occurs at the end of the pipe. This is because the properties of the medium (air) inside depends partially on the shape of the pipe. Where the pipe abruptly ends, a condition is imposed on the two traveling waves: *Total wave amplitude = a maximum*. If the pipe is open at both ends, a wave maximum occurs at both ends. If one end is closed, the amplitude equals zero at that end, and is a maximum at the open end. In this case, partial reflections will occur and standing waves will appear.

What are the Rules for Adding Two Waves? What happens when two waves travel together? To learn this we need rules from Nature for adding together the amplitudes of the waves. These rules are simple: you always add amplitudes where two positive crests or two negative crests occur together, and you subtract them if negative and positive occur together. In between, add the waves and follow the rules of algebra with sine and cosine expressions. The technical term for the phenomenon of waves reinforcing or cancelling each other is *interference*.

Interference of Dissimilar Waves Produces Beats. When two waves travel together there are two ways in which their amplitudes may join. The first of these is an *interference in time*. An example is two sources of sound which have slightly different frequencies which we listen to at the same time. When the waves arrive with the crests together so that the amplitudes add; we hear a sound peak. When a positive crest of one and a negative crest of the other are together, the amplitudes cancel and we hear nothing. The rising and falling of sound is called beats. It is an interference in time.

Repeated Reflections Create a Standing Wave. The second way is an *interference in space* which occurs when two waves of the same frequency are confined in the same space such as in a trombone pipe or on a violin string. The wave is confined because there are reflections at both ends of the string or pipe. One wave is traveling towards an end. Its reflection becomes a wave traveling in the other direction. If the crests of the two waves always occur together at the same place, if both waves have identical frequencies, and if the reflections occur continually, then the two waves appear motionless and a *standing* transverse *wave* is seen. The violin string moves to and fro. This is the normal situation in a musical instrument.

Resonance is a Love-affair between Two Instruments. If two standing waves, say one on a violin and the other on the piano, have the same frequency, then sound energy can be transferred between the instruments. You can try it. Place the violin near the piano and strike a piano key corresponding to one of the tuned strings. Stop the piano note and you will hear that the violin has picked it up! To describe this, we say that the strings of the two instruments are in *resonance* with each other.

You can also try this just using the piano by striking a bass register note, say D, and at the same time holding the damper off a higher D note. The higher D will resonate with the lower note and absorb some energy. You can hear it by taking your finger off the bass D note while holding the higher D. The high note will remain.

A resonance condition can occur in other situations, too. People who say, "My ears were ringing!" were probably in resonance with a component of a loud sound. Lovers may claim that their thoughts and feelings resonate with each other! The yellow light emitted from heated sodium atoms will be absorbed resonantly by a different layer of sodium vapor placed in the path of the light. A tuned device is often termed a *"resonance* or *resonator"* of a particular frequency if its dimensions match that frequency.

Intensity of a Wave is Proportional to the Square of the Amplitude

Light waves, as well as other types, are able to carry energy and momentum from one oscillating atom to another. When this occurs in a light beam, measurements show that the intensity or power of the beam is proportional to the square of the amplitude of the electric field. That is

Power α E*E watts per square area

This result can be obtained by solving Maxwell's Equations which describe the total effect of many separate quantum exchanges. The same intensity relation is true for other waves, such as sound.

Electromagnetic Waves are Special

The waves of light, radio, x-rays and other members of the electromagnetic spectrum stand apart from sound, and other mechanical waves. They are unique as well as different. Their speed of propagation c, is regarded as a fundamental constant of Nature, unlike the speeds of mechanical waves which vary widely and depend on different mechanical properties of the medium in which they travel.

Nature appears to have chosen these waves as the messengers which bear the laws of Nature. Only e-m waves are described by the fundamental equation:

energy = mc^2 = hf

The energy **mc^2** implies the law of relativity, and the energy **hf** implies the laws of quantum theory. Laws and calculations of the waves yield precise matches with measurements. Still, we don't fully understand them. Especially, the notion of a photon is fraught with contradictions and space, their medium, has hardly been investigated.

Also unique, and perhaps just as fundamental, are the "matter waves" of quantum theory. But our knowledge of them is even less. We know they are scalar waves and we can calculate their behavior precisely. Their frequency is proportional to the energy of a particle and their wavelength is inversely proportional to the momentum of the particle. Except for the proposal in Chapter 12, there are no physical models of their structure, and we know almost nothing about the media.

In contrast, we feel that we thoroughly understand a *sound wave* because its medium of propagation is the air in which it travels from a source to the listener. Sound is a study of the motion of gases and can be completely understood in terms of Newton's laws, we imagine. The propagation from one place to another is merely a consequence of mechanics and the properties of the air media. If it propagates through a liquid or a solid, it depends on their mechanical properties. Familiarity gives us confidence and mitigates our skepticism. But upon close examination, sound waves are found to be composed of atoms undergoing quantum waves. So our confidence is an illusion!

You Never Know a Wave Until You See it, Hear it, or Find it!

In the discussion above, an extremely important but subtle assumption has been made. We assume that amplitudes are able to reinforce each other simply because we ourselves are able to hear sound waves or see light fringes. In fact we haven't thought about how Nature combines or senses wave amplitudes. Reflect for a while about this. People see or hear the interaction of waves using the automatic sensors built into our heads. From childhood we learn about light and musical sounds by experiencing them with our own senses, so to assume this process always exists is natural. However, for waves outside of human sensory experience this assumption cannot be made. Instead, we must understand the process of interaction between waves, and the means of detecting them. We need to study the mutual detection process involving energy exchange which occurs when atoms and particles interact with each other.

In fact, the act of *detection* is the act of being an *observer*, a concept which is central to relativity and quantum theory; things we would like to understand better. This strikes at the heart of the problem of understanding space. We will see in Chapter 12 that space may possess the property responsible for the interaction and combination of different waves.

For electromagnetic waves in electronic apparatus the detection process depends on a *non-linearity* of the media in which the waves travel. This is the subject of the next section.

Waves Travel Peacefully when the Media are Linear

Imagine a device which has an output amplitude which is proportional to an input:

$$A(out) \propto A(in).$$

If this is true the behavior of the devices is termed *linear*. The device might be an object whose acceleration is proportional to the force applied, or an electronic circuit element in which the current flow is proportional to a voltage applied, or wave media.

When the wave media are linear, the waves do not interact. For example, if you throw two stones into a lake, each will create a circle of expanding waves that intersect each other. As you watch, each set of circles grows larger without interference by the other. The surface of the lake is a linear wave medium.

A stretched string of uniform size and tension is a linear medium. Waves traveling along it can meet waves traveling in the other direction and both will pass through each other without disturbance. Billions of electromagnetic signals travel through space, constantly intersecting each other, yet never interfering or mixing with each other. Space is clearly a linear medium for most e-m waves.

A Non-linear Medium Transfers Information Between Waves. We are more interested in *non-linear* devices where output is *not* proportional to the input. That is, a graph of input versus output will not be a straight line through the origin. Non-linear devices are important to energy exchange. Consider what happens if we apply a pure sine wave tone to the input of a non-linear device. First, the output will not be a sine wave; instead it will be distorted, and harmonics of the sine wave are produced. Second, if the input should be two or more waves, the output will not only contain harmonics of the two pure inputs, but each wave will contain components of the other. That is:

If the INPUT $= A_1 \cos(f_1 t) + A_2 \cos(f_2 t),$

then you get:

OUTPUT $= k A_1 A_2 \{\cos(f_1 + f_2) + \cos(f_1 - f_2)\} +$ input $+$ harmonics

The two signals mix; that is, the non-linear device has created *modulation* of one signal by the other. Each signal wherever it goes carries information about the other signal. Notice that this effect is proportional not only to the degree of non-linearity **k**, but also to the product of the two amplitudes.

There are many real situations where non-linear devices occur. Our ears are non-linear which accounts for our ability to hear sum and difference frequencies (beats) when two musical instruments are played near each other.

Radio transmitters incorporate a non-linear device, called a *modulator*, to impress the audio signal to be transmitted onto the radio carrier wave. The radio receiver also includes a non-linear device termed a *detector*, which extracts the desired audio signal from the carrier. On the other hand, hi-fi sound equipment is carefully designed to be as linear as possible to avoid distortion. Laser light sources are so intense that they cause the wave-transmission properties of glass to become non-linear so that when red light passes through, its second harmonic comes out as ultraviolet light.

The detection of an e-m wave traveling in space may also be a non-linear event. When an e-m wave encounters a charge, say an electron, we observe that the wave affects the electron, or vice versa, and can transfer energy to it. How does this happen? Does the electron create a bit of non-linear space? Or does the electron have a special property of its own? At this time, science has no answers for these questions. Chapter 12 makes a proposal to explain this.

The Magic of Complex Numbers can be Applied to Sinusoidal Waves

In Chapter 5 we saw that an exponential function could be made equal to the complex sum of a sine and a cosine function. This fact can be turned into a neat mathematical trick to make the algebra of waves much easier. Exponentials are easy to manipulate.

All you do is to let the amplitude of the wave equal an exponential function: $e^{\{2\pi i(ft - x/\lambda)\}}$. The real part of this exponential is a sine function. Since the real and imaginary parts never mix with each other during manipulations, you may do all the algebra you want using the exponential function. Then at the end, just use the *real* part of it. This yields the same result as if you used the sine function all along.

Frequently you need to find the square of the amplitude of a wave. This is also easy using an exponential. All you do is multiply the complex amplitude by itself except that every "i" is replaced by a "-i". This result is often written $[A]^2 = A^*A$, where the A^* is the amplitude with "i"s replaced by "-i"s.

In the next chapter, you will see the application of wave ideas to the mysterious quantum theory, a fundamental law of the universe which intimately involves waves.

"Anything that is not forbidden is allowed."

— V. Ginsberg

CHAPTER 9

Quantum Mechanics

$$\lambda = \frac{h}{p}$$

Interference of Waves
The Hydrogen Atom Waves are Like Sand Patterns on a Drumhead
History of Quantum Mechanics
The Amplitude Ψ Tells you Where the Particles Are
The Marriage of Quantum and Relativity
The Trouble with Quantum Electrodynamics
Interference of Quantum Waves
The Uncertainty Principal: You Can't Have it Both Ways
The Ultimate Paradox: Bell's Theorem

CHAPTER 9
Quantum Mechanics

Quantum mechanics is the theory of how matter, light, and energy behave on the scale of atomic events. This is the realm which we cannot observe with our human senses. Nearly all the information about quantum behavior has been obtained with measuring instruments that detect atomic sized events and provide data which we then interpret in terms of mental models and mathematical relationships. Although we cannot see and feel the models directly, we have learned to trust most of the ideas because the mathematics predicted the outcome of quantum experiments beforehand, and usually, the experiments showed that the predictions were correct. Of course, when the predictions failed, the theory was discarded. What remains is consistent and accurate.

In general, quantum behavior boils down to the fact that all particles tend to behave as though they were a wave whose wavelength depends on the momentum of the particle:

wavelength = h/momentum,

where **h** is Planck's constant (6.6×10^{-34} joule-second). If you want to calculate a particle's behavior, you must deal *not* with Newton's laws of motion, but with the properties of waves and the conservation of energy (frequency). The rules for calculating waves are almost the same as the rules for ordinary light waves. Then, mechanical properties are found by using the rule that a particle, whatever it is, takes on the energy and momentum associated with the waves, according to the rules $E = hf$ and $p = h/\lambda$, where λ is the wavelength.

Interference of Waves

Quantum waves display interference and reinforcement like light, and in cases where the waves repeat the same path of travel, the waves can only exist if the crests and valleys of each wave passage occur in the same place as the previous wave. Otherwise, the waves will cancel each other out. This leads to a situation where only certain resonating wavelengths may occur. In quantum slang, we say they are "allowed" and all other waves are forbidden. This is closely analogous to the allowed harmonic waves in a musical pipe or on a string. Indeed, the arrangement of quantum waves among particles is not unlike those on a drum head, in an organ pipe, or the body of a violin.

Perhaps the most important example of *allowed* wave states is the hydrogen atom shown in Figure 9-1. Around the proton, the electron waves form 3-dimensional patterns of standing waves with a single frequency, so only that energy (frequency) is allowed. The normal atom is the simplest pattern of lowest energy, a symmetrical spherical wave around the proton. More complicated lobed patterns occur at higher energies. These plots are solutions of Schroedinger's wave equation as given in textbooks of QM. For clarity I have condensed all the natural constants and fixed factors into one coefficient, so that you can more easily see the dependence on the variables, r, Ω, and ø.

The Hydrogen Atom Waves are Like Sand Patterns on a Drumhead

Figure 9-1 uses spherical coordinates similar to the radius, latitude, and longitude of the Earth. These are convenient because the atom has spherical symmetry. If you carefully examine the figure, you will see that wave patterns occur in a radial direction and circularly around the sphere. To understand the wave patterns, recall what happens when sound waves are confined inside some hollow object like a box or a drum. Standing waves occur perpendicular to the walls because waves reflect back and forth between them. In a drum one or more standing waves (different harmonics) resonant between the drum heads. Others can travel circularly around the cylinder. These standing waves can be made visible by sprinkling sand on the drum head which will then display the wave pattern.

The standing waves which are *allowed* depend on the box or drum dimensions and shape. The allowed rule is that waves which travel the same path must repeat the same amplitude pattern, otherwise successive waves will cancel each other.

It is evident that electron waves are traveling around the proton and inward and outward from the center. The allowed patterns in the circular direction are easy to visualize, being quite similar to those which would occur for sound waves traveling inside a drum. The radial waves are a little mysterious because they are not reflecting between two walls, unless you are willing to imagine one wall at $r = 0$ and another at $r = \infty$. In fact that is the situation!

The integers **N**, **L**, and **M** correspond to the octaves of a piano, where one speaks of a 1st harmonic, 2nd harmonic, 3rd... etc. On the piano they refer to the integral number of wave crests which fit onto a piano string. As the integers become higher, it means that more wavelengths are being fitted into the space where the waves travel. Obviously only an integral number of waves are allowed, otherwise the uneven waves would cancel each other. Higher numbers mean greater frequency (greater energy) of the electron wave state.

CHAPTER 9/*Quantum Mechanics*

There are no Electron Orbits! Whoever started the notion of electrons traveling around the nucleus like planets made a terrible blunder! If you have learned such an idea, discard it immediately. Instead, all calculations and all experiments show that *no* satellite-like orbital motion exists in the normal atom. Instead, there are standing wave patterns. For example, see the case of N = 1 in Figure 9-1 where the standing wave pattern is entirely spherical. The center of the electron pattern is also the center of the proton pattern. This is the normal situation of the H atoms in the universe; they have spherical symmetry, not orbits.

If you are interested in studying these patterns carefully, which I would suggest since standing quantum waves are the basic structures of the matter of the universe, then you have to memorize the spherical coordinate variables θ (or Ω), φ, and **r**. Here is a trick. First, the radius (**r**) is easy. Then, you notice that *theta* (θ) looks like a little sphere with a latitude line on it; so θ is the latitude variable. You also notice that phi (φ) looks like a sphere with a longitude line on it so φ is the longitude variable.

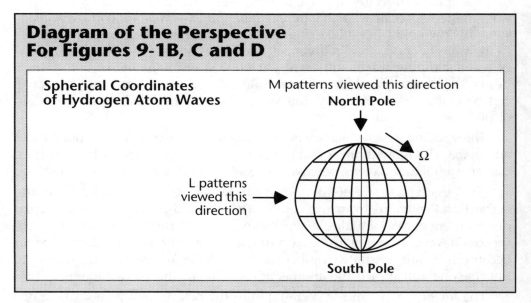

Figure 9-1A. The hydrogen atom wave functions.
The Figures 9-1B, 1C, and 1D show the probability density of electron waves surrounding a proton in the hydrogen atom. The total wave is the product of one of each kind: TOTAL = BxCxD. The normal atom has N = 1, L = 0, and M = 0 which are the upper pattern in each figure. The waves are in three dimensions, so the total wave is a combination of three kinds of standing waves, all resonating together. The first kind, with N quantum integers, are waves moving radially outward from the center where the proton is located. The other kinds, M and L quantum integers, are standing waves around circles of symmetry in the sphere.
A, A', A", B, B', R, etc. are numerical constants that make the total probability =100%.

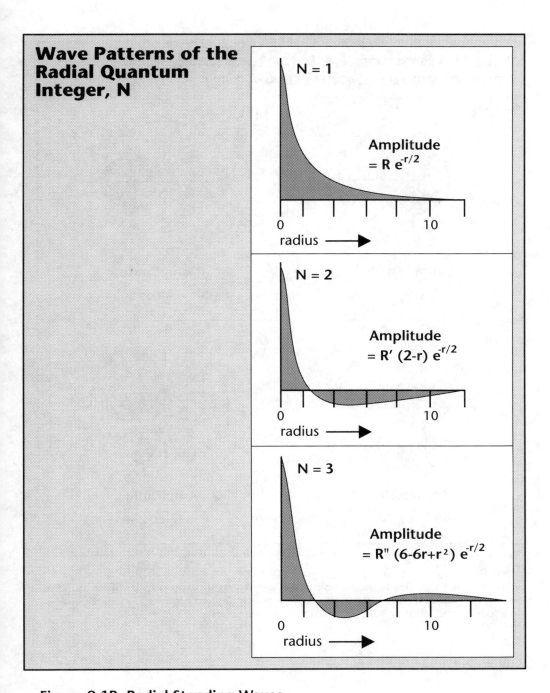

Figure 9-1B. Radial Standing Waves.
In each figure the radial wave amplitude is shown for a different value of the quantum integer N. Each added integer adds one more crest, like standing waves on a violin string.

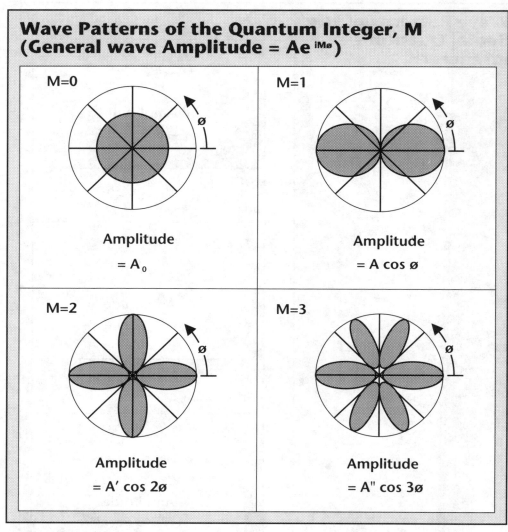

Figure 9-1C. Equatorial waves.
These figures look down upon a pole of the sphere and show amplitudes of the standing waves rotating around the center of the equator which depend on the longitudinal angle φ. Each added integer adds another pair of standing wave lobes.

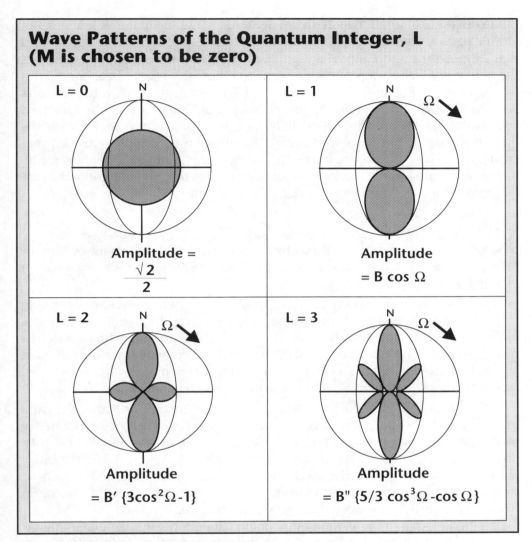

Figure 9-1D. Polar standing waves.
These figures look towards the equator of the sphere and show waves traveling between the poles. The amplitude equations are shown as Associated Lengendre Functions which depend on the latitude cosΩ, and the quantum integers L and M. Each added integer of L adds another pair of standing wave lobes.

Connection with the Human Realm. All solid matter is made up of atomic particles which have to be dealt with using quantum rules. But we find that as more and more atoms are grouped together to form objects, the more the behavior of the group of atoms begin to follow the human scale rules with which we are familiar. Thus, the quantum laws gradually turn into Newton's laws, as the size of the group increases. But we can't forget that the quantum laws and behavior are the real basis of all mechanics. What lies beneath the quantum laws? What is the unseen machinery or model of the particles that create these laws? We do not know. But we do know that particles are not tiny lumps, not scaled-down specks of sand, nor orbiting things. Their properties are quantum properties most easily understood as standing waves.

History of Quantum Mechanics

Knowledge of the quantum realm was accumulated gradually during the first half of this century. It has been an intriguing detective adventure for several decades. The quest continues, since there is still no general explanation of why it happens this way.

The first clue was noticed by Max Planck, about 1900, when he was trying to puzzle out the reasons for the observed spectrum of the light from a hot object. He discovered that he obtained the right equation which matched measurements if he assumed that light was exchanged between atoms in fixed amounts whose energy was proportional to light frequency, **E = hf**. These fixed amounts he termed "quanta" which is how the name began.

In 1904, Albert Einstein proposed another relationship for Planck's quanta based upon the experimental observation that a tiny current of electricity would pass between two electrodes in a vacuum if light was allowed to strike one of them. This was to become known as the *photoelectric effect*. The electricity always flowed from a negative electrode to a positive electrode, so he deduced that negative electrons were involved. The experimenters found a minimum frequency at which the effect began. As the light color was changed towards higher frequencies (more towards the violet), the electrons had higher energy. At lower frequencies (towards red) there was no effect. Different metals displayed different frequencies (colors) where the effect began. The intensity of the light always increased the intensity of the electron flow. Typical such measurements are shown in Figure 9-2.

Einstein deduced that the incident light was divided into discrete amounts, as proposed by Planck, and that each quantum had an energy **hf**. Part of this energy was used by the electron to escape from the metal, and the remainder became kinetic energy of the ejected electron, which is seen as a potential energy or voltage between the electrodes. Einstein's photoelectric equation is:

hf = metal work energy + kinetic energy

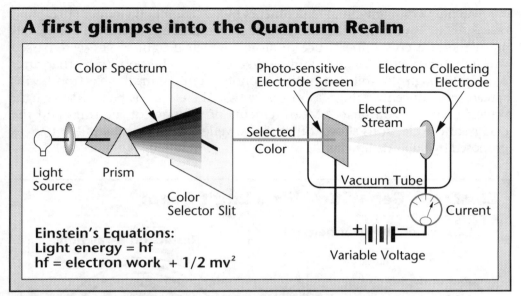

FIGURE 9-2. How the photo-electric effect is measured
Light goes through the prism and is separated into its colors. One color is selected by a slit. That ray knocks out an electron from the photo sensitive electrode. Its energy is measured by the voltage needed to repel it.

Einstein (1905) deduced that the light communicates an energy exchange to the electrode which it strikes, resulting in the ejection of an electron from the photosensitive metal surface. The energy of the light is delivered in discrete quanta of size: energy = hf. This amount of energy is found to be equal to the work required to eject the electron, plus the kinetic energy of the moving ejected electron.

He realized that if the energy of the incident light **hf**, did not exceed the metal work energy, then no electrons could escape and the current would be zero as observed.

Then he deduced that the electron's kinetic energy must equal the voltage energy (**V**) that appears if you place a sensitive voltmeter between the electrodes,

$$1/2\, mv^2 = eV$$

where **e** is the charge of the electron. This was found to be true.

The quanta of light energy soon became known as **photons** and some persons regard them as particles, although there is no experimental evidence of particle-like properties except that the energy exchanged always begins at some small place such as an excited atom, and ends up at another small place such as an atom of a detector.

The next clue came from experiments by Davisson and Germer (1927) and others, in which beams of electrons were shot through thin metal foils (metals are made of a crystalline lattice of atoms, which if light waves were passed through them would cause rainbow-like effects). They observed that after passing through the foil and hitting a photographic film, the electron beams spread into concentric rings, as in Figure 9-3. The electrons behaved like light waves; they were diffracted! Measurements of the sizes of the rings and the energy of the electrons showed that the wave length λ was always **λ = h/p**, as proposed by Louis de Broglie.

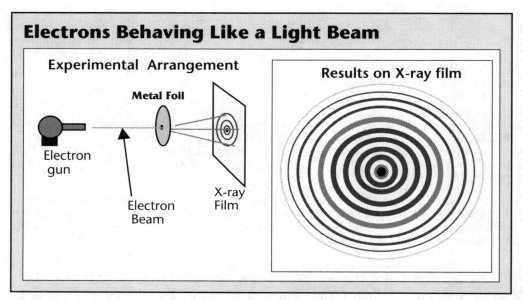

FIGURE 9-3. Diffraction of an electron beam
The electron beam passes through the metal foil which acts like a small aperature grating. The electron beam has wave properties, so the pattern on the film shows wave interference patterns.

Fundamental Rules: These two observations, **E=hf** and **λ=h/p**, are fundamental measured facts about quantum theory. We don't know the cause of these two relations nor do we have any model of particles which predict them; they are just measured experimental facts. From these two observations, many mathematical ideas have been deduced: some wrong, some right. The correct ideas were sifted from the wrong ones by making a test of the mathematical predictions of the various theories, using a new experiment.

RESEARCH GOAL 9-1: *What is the origin of the QM relations:*
E = hf and λ = h/p?

Wave Equations tell you How the Waves Behave. It was soon realized that a means had to be found to calculate the location, amplitude, and frequency of the waves if the new quantum effects were to be understood and be useful in the analysis of physical problems. The means which have been developed boil down to the following: Whenever you are concerned about the behavior of a particle in the quantum realm, the typical way to begin is to write an equation that will describe the waves of the particle. This can be done, by using the above two basic relations to obtain the frequency **f**, and wavelength λ, of the particle.

For example, a sinusoidal wave traveling along an x axis can be described by an amplitude:

amplitude, $\Psi = A \sin 2\pi(\mathbf{ft} - \mathbf{x}/\lambda)$,   or, as an exponential

$$\Psi = A \exp i2\pi(\mathbf{ft} - \mathbf{x}/\lambda)$$

where **A** is a constant. This equation is just an ordinary rippling wave; it doesn't describe anything in particular. But, by substituting the two fundamental quantum relations, $E = hf$ and $\lambda = h/p$, it can be rewritten:

amplitude, $\Psi = A \exp\{i2\pi(Et/h - xp/h)\}$

Now, the equation describes the wave of something which has values of energy E, and momentum p, the same as the electron we are studying.

Half a century ago, Erwin Schroedinger noticed that if you use the rules for differentiation on the above equation, you can immediately find that

$\partial \Psi/\partial t = \{2\pi i E/h\} \Psi$,   or $E \Psi = i(h/2\pi) \partial \Psi/\partial t$

and that

$\partial \Psi/\partial x = -(p/2\pi)/h \Psi$,   or $p \Psi = -i(h/2\pi) \partial \Psi/\partial x$

These results suggested that:

E is equivalent to the operation of using $- i(h/2\pi) \times \partial$(wave function)$/ \partial t$ and $\mathbf{p_x}$ is equivalent to the operation of using $- i(h/2\pi) \times \partial$(wave function)$/\partial x$.

The variables **E** and $\mathbf{p_x}$ are replaced by operators of differentiation, multiplied by constants. The same is true for momentum in the other two directions $\mathbf{p_y}$, and $\mathbf{p_z}$. Since momentum is a vector obeying the pythagorean rule, the total momentum is $p^2 = p_x^2 + p_y^2 + p_z^2$. Since the **p** operator is squared, it has to be applied twice.

From these observations, a method has been invented to obtain a wave equation by simply writing out an expression for the total energy of the particle times a wave function. Then wherever energy **E** or momentum **p** occurs, the above equivalent operators are substituted. This produces a wave equation containing differential operators.

For example, in order to write a wave equation for a particle satisfying both QM and relativity, the expression for the energy of a moving particle is used to operate on Ψ:

$$E^2 \Psi + p^2 c^2 \Psi = m^2 c^4 \Psi \qquad (9\text{-}1)$$

Then **E** and **p** are replaced with their equivalents to obtain a differential equation which, when solved, provides a description of the wave amplitudes Ψ of the particle. Sometimes the solution will resemble the waves on musical strings, in pipes, or drumheads; that is, only certain tones (frequencies) exist. These *allowed* frequencies correspond to fixed energy levels E_n, for the particles. Both the wave amplitudes Ψ and the E_n can be calculated.

The above scheme is not a rigorous mathematical process — it is just an educated guess that works most of the time. The process is not without pitfalls. There are often many choices. Not all of them yield meaningful results, nor are all solvable.

The ultimate proof of the quantum theory method has been that the results agree with experiments when the right choices are found. In particular, all the energy levels of the hydrogen atom have been calculated and they agree within experimental error. The accuracy obtained has been amazing. Complicated atoms have not been calculated exactly, because the complex equations that result from many electrons are too difficult to solve.

The Amplitude Ψ Tells You Where the Particles Are

How should we regard the function Ψ (x,y,z)? It is just an equation which finds an ampltitude number at various points in space depending on the position variables **x**, **y**, and **z**. Mathematical analysis and experimental trial and error show that where the amplitude is large, the probability of finding the described particle is large. This is the accepted meaning of Ψ.

Since probability must always be a positive number, it won't do to think of Ψ, sometimes a negative number, as a probability. Instead, logic and trials show that the probability is given by its square,

Probability (x,y,z) = |Ψ(x,y,z)|²

which is always a positive number. It is interesting that the square must be used, because this frequently occurs in other branches of physics where the realities of Nature: such as light intensity, kinetic energy, and liquid flow are given by the squares of an amplitude.

If you are already familiar with statistics, you will know that the total of a probability cannot be more than 100%; one electron cannot have a probability of existing 101%. So, there is a restriction,

The sum of the values of $|\Psi|^2$ everywhere = 100%

or, in mathematical notation,

$$\int^\infty |\Psi(x,y,z)|^2 = 1$$

This requirement is met by calculating a suitable coefficient **k**, so that $\int |k\Psi|^2 = 1$. Then the function **k**Ψ is said to be *normalized*.

The Marriage of Quantum and Relativity

No correct theory of the *motion* of matter can ignore the principles of relativity, because the only velocity between two objects is the relative velocity. For this reason, it was soon realized that quantum mechanics had to be joined with relativity to provide a correct description of the physical world. This is accomplished using mathematical methods termed *quantum electrodynamics* (QED).

The essential feature of QED is to make a differential wave equation beginning from the relativistic energy equation (9-1). Then **p** and **E** are replaced by their equivalent operators. This produces,

$$-(h/2\pi)^2 \partial^2\Psi/\partial t^2 + (h/2\pi)^2 c^2 \partial^2\Psi/\partial r^2 = m^2 c^4 \Psi \qquad (9\text{-}2)$$

which is called Schroedinger's relativistic wave equation. This equation when solved yields waves with both the required relativistic and quantum properties, but it does not display a spin value to match the electron. This difficulty arises partly because the terms of energy **E** and momentum **p** are squares, rather than first order.

Dirac's Serendipity got rid of the Squares. Paul Dirac pondered over equation (9-2) and cleverly discovered (1933) that if you take the square root of both sides of (9-1) by rewriting it using matrix quantities **a**, **b**, and **i**, you can get rid of the squared terms as follows,

$$E \underline{a} \Psi + p c \underline{b} \Psi = \underline{i} m c^2 \Psi \qquad (9\text{-}3)$$

CHAPTER 9/*Quantum Mechanics*

The solutions take on matrix properties too, so that Ψ, **E**, and **p** have four components instead of one component if you use ordinary algebra. Amazingly, these extra components provided the two spin values of the electron and predicted two kinds of electron, which were thought to be the electron and the positron (discovered a few years later). Even more convincing was that the solutions yielded accurate values of spin and magnetic moment. His result appears both magical and serendipitous. There is no logical explanation of why it has worked out so well, but the accuracy of the many and varied predictions cannot be denied. The Dirac equation has become the backbone of the theory of the behavior of electrons and photons, termed *quantum electrodynamics*.

The mathematical results of the Dirac equation are very important and yield many added clues to the ultimate nature of space and particles. It is a fertile subject to study for those persons who like to wade through the ponderous mathematics.

The Trouble with Quantum Electrodynamics

There are serious logical problems with the quantum theory when it is appliedto the electron or other point-like particles. One of the important terms in the mathematics of QED is the "self-energy" of a charged particle, such as an electron, which has an electrical potential energy assumed to be given by $V = e^2/r$. The self energy of a charged particle depends on the radius **r** according to $1/r$. Thus if the particle size is shrunk down to a point $r \to 0$, the self energy goes to infinity. Besides being impossible, the equations become useless. This is a problem.

To avoid the infinity dilemma, one is tempted to abandon the idea of a point particle. But relativity will not allow this, as seen from the following argument. If a particle is *elementary*, it must react as a unit. However if it has a finite size and an electromagnetic signal should arrive at one side, the other side must simultaneously know of the arrival of the signal in order to react as a unit. But this implies that the signal travels with infinite speed which is prohibited by relativity. The only way out is to have a point particle. (Or no particle at all, if you could find a way to represent mass and charge without it.)

The theoretical physics community has resolved this impasse by accepting the point particle and inventing a process called *renormalization* (Feynman, Schwinger, and Tomanaga) in which an infinite energy is subtracted from the infinite energy to obtain a finite number. Unfortunately, the result could be *any* finite number. In the accepted chosen process, the result is equivalent to assuming that the electron's charge force becomes zero inside a distance of roughly $R = e^2/mc^2$, the classical radius of the electron. There is no explanation for this coincidence; it merely happens that way.

Acceptance of renormalization is made palatable by the exceptionally accurate agreement of the calculations with experimental measurements, especially two results:

1) The magnetic moment of the electron differs slightly from the value calculated by the plain Dirac theory. Using renormalization, the agreement is excellent.

2) Measurements (Lamb, 1947) of very fine energy levels of the hydrogen spectrum, due to the tiny interaction of the electron spin with the proton spin differed slightly from theory. But after renormalization, the agreement was perfect within experimental error, which was only 10^{-8}!

The very close agreement of these two observations gave many people a great deal of confidence in both quantum theory and renormalization. While the mathematics of renormalization is not demonstrably consistent, this agreement gives many the feeling that beneath it, in some undiscovered way, it is a bona fide theory.

Nobel laureate Paul Dirac who developed much of QED, did *not* feel comfortable. He writes, (1937), "This is just not sensible mathematics. Sensible mathematics involves neglecting a quantity when it turns out to be small — not neglecting it because it is infinitely large and you do not want it! Of course the proper inference is that the basic equations are not right. There must be some drastic changes introduced into them so that no infinities occur in the theory at all."

The space resonance theory proposed in Chapter 12 provides a reason why renormalization works; the resonance's standing waves have a small amplitude inside $R = e^2/mc^2$. Is this Dirac's drastic change?

Pros and Cons of the Particle Concept. Throughout history, the period of development of the quantum theory, and at the present, there is a strongly believed concept that "particles" exist. In QM they exist somewhere in the environs of the wave function of an electron.

There should be considerable doubt that the particle is a valid concept, because there are few scientific facts to support it. Consider the evidence:

1) All experiments to probe a central structure of the electron have been negative.

2) No QM theory exists that predicts a size for the electron, a mass, nor a charge. Further, there is no theory that quantifies the particle in a meaningful calculation. This implies that QM actually has no need

of a particle concept because all the calculations are the same whether or not you believe in particles.

3) The substantiality of "mass" is doubtful because it can always be converted to electromagnetic energy, which has no particle properties.

On the other hand, there are substantial emotional and historical reasons for maintaining the concept:

1) The approximation that mass can be located at a precise point is extremely convenient for engineering calculation. The particle concept is widely found in technical books, creating a bias for its preservation.

2) Because of the construction of our eyes and of microscopes, any object smaller than wavelengths of light appears to be a point. This point concept of a particle has become fixed in tradition and our emotions.

It can be reasonably concluded that the quantum theory has no need of the particle — only humans have the emotional need. Whether or not particles actually exist becomes a philosophical question. But we can't forget that mass and charge are really there, associated with the particle. If we throw out the particle, we must still keep the charge and mass. Can this be done using a wave?

The Size of Quantum Waves. The waves in quantum theory are especially ephemeral because we never see them or their effects as clearly as we do a water wave, a sound wave, or a one-meter TV wave. We can observe those waves directly with our eyes. (The TV wavelength can be noticed when an airplane flies between the transmitter and our home — it causes the picture to flutter as the reflected wave and direct wave alternately reinforce and cancel each other.) Even a wavelength of light can be seen if you are fond of playing with optical gratings.

Quantum waves, on the other hand, are very small. Calculate the wavelength of an electron in the TV picture tube. If its velocity is 86% of **c**, then the relativistic factor $(1 - v^2/c^2) = 0.5$ so its mass **m**, has doubled. The momentum of the electron is $\mathbf{p = mv = 2 \times m \times 0.86 \times c}$. The deBroglie quantum wavelength is then $\lambda = \mathbf{h/mv} = 6.6 \times 10^{-34}$ joule-second$/(2 \times 9.1 \times 10^{-31}$ kg $\times 0.86 \times 3 \times 10^8$ meter per second$) = 1.4 \times 10^{-12}$ meter. This is a million times smaller than visible green light wavelength, which is about 0.5×10^{-6} meter. λ is very tiny! Notice that the deBroglie wavelength is inversely proportional to the momentum. So, heavier particles such as a proton, or faster electrons, have wavelengths even smaller.

There is no existing apparatus such as a radio receiver which can directly detect such small wavelengths, so we humans are deprived of the opportunity to have personal experience with them. Nevertheless, the presence of the wave results in the formation of atoms, composed of protons and electrons. The wave is an important aspect of semi-conductors, such as the microchips of the computer on which this page is being typed. The wave is really there!

Interference of Quantum Waves

Two quantum waves can interfere with each other, just like light waves or TV waves, combining so as to either reinforce their amplitudes or cancel them. A common way of creating two identical waves is to allow one wave to pass through two holes in a plate. Then the two waves are detected at a screen on the other side. This is illustrated in Figure 9-4.

The pattern which is detected at the screen (moveable detector), has maxima and minima due to interference produced as follows: At the center, the distance from hole **A** is the same as to hole **B**, so both of the waves have the same phase. They are alike, and reinforce each other to create the maxima observed. At some equal distance, above and below the center, wave **A** and

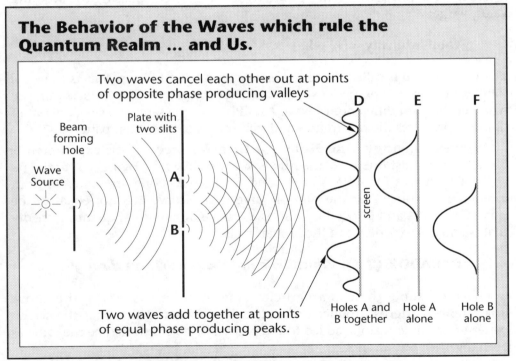

Figure 9-4. Wave amplitudes can interfere with each other.
The amplitudes of waves from holes A and B create an interference pattern at screen D where the amplitudes either combine or cancel each other, producing peaks and nulls at a detector. Both light waves and quantum waves behave as shown here.

CHAPTER 9/*Quantum Mechanics*

Wave **B** have unequal distances to travel and are 180° out of phase. They are opposite, and cancel each other to create the minima observed. This behavior is repeated moving further away from the center in both directions, resulting in more minima and maxima.

There are two quantum rules to calculate wave interference:

RULE 1: Add the amplitudes of the interfering waves everywhere.

Then, to obtain the intensity measured at the screen:

RULE 2: Square the absolute value of the total amplitude.

For example, using exponential waves of frequency **w**, we sum the two intensities,

$$I_A + I_B = A\,e^{i(wt+d)} + B\,e^{i(wt+d)}$$

where **d** is the phase difference of the two waves. Then squaring the absolute value, we get

$$\text{total intensity} = [I_A + I_B]^2 = A^2 + 2AB\cos d + B^2$$

The middle term is called the "interference term." It adds or subtracts from the total according to whether **cos d** is positive or negative. That is, according to whether the path difference of wave **A** and **B** have the same phase or are out of phase. It produces the maximums and minimums of the screen pattern.

Waves Through One Hole: No Interference. If this experiment is repeated using just one hole, the results are also shown in Figure 9-4, to the far right. The curves for hole **A** alone and for hole **B** alone are alike, without interference. Notice that the sum of curve **A** and curve **B** is not equal to the curve for hole **A** and **B** together! This is a unique difference of quantum theory that we do not experience in the human realm.

PARADOX (9-1): *Which hole did the particle go through?*

If you assume there is a particle somewhere in the waves, the above interference patterns produce a paradox. You reason at first that the screen pattern from hole **A** alone added to the pattern of hole **B** should be the same as both together. But it is not.

Then, you reluctantly agree that the waves could be a sort of "guiding influence" for the particles. And if you are a meticulous logician, you might start with the following,

ASSUMPTION: *The particle goes through either hole A or B.*

Then you discover that the number of particles which go through **A** plus **B** separately is not equal to the total number when **A** and **B** are open together. So the assumption appears to be wrong, or perhaps the waves affect the particles as they enter the holes! How can the mere presence of one hole affect what enters the other hole?

However, if this experiment is done using ordinary light or microwaves, the results are still the same, and we see no paradox, since there is no particle to worry about. So we suddenly think, "Aha, the answer is simple! Forget the particle! There is no particle!" But then we sadly find another problem, because electrons or other particles, have charge, mass, and spin, so we are stuck without a way to explain how these properties are carried along by just the wave.

Nevertheless, ignoring the particle is a good idea, because the theory of QM and the calculations that come out of it do not depend on the notion of a particle. So in real fact, you can ignore the particle most of the time and you won't go wrong. You just have to keep in mind that the charge, spin, and mass of the particle are still there. If you can find a different way to account for them, the particle is not needed and there is no paradox.

Figure 9-5. Can two particles occupy the same space?

In the human realm this is impossible. But in the quantum realm of atoms, it often happens. In the normal hydrogen atom, the proton and electron waves are spherically symmetrical around the same center. Yes, they occupy the same space. See figure 9-1 (N = 1, L = 0, M = 0).

The enigmas are there and yet there seem to be no flaws in the method. If we try to claim that something is wrong with the basic theory, we have to account for the fantastically accurate results of quantum theory in predicting the measurements of interacting photons, electrons, and simple atoms. This leaves limited room for new ideas.

But in Chapter 12, a structure for particles formed out of space is proposed, which appears to overcome these paradoxes. A way is found for the waves alone to carry the charge and mass and there is no particle.

The Uncertainty Principle: You Can't Have it Both Ways!

Early in the history of QM, Werner Heisenberg proposed a principle which states the limitations on the accuracy of physical measurements. He stated that Nature imposes a minimum value, Plank's constant **h**, for the product of the errors when measuring position, **Δx**, and momentum, **Δp**, when both are measured together. This is usually written

$$\Delta x \, \Delta p \approx h$$

This idea created a great deal of controversy, since it says that nothing is certain. There is always an error in measurements today and these errors will grow larger tomorrow. Heisenberg's principle expanded into discussions of the nature of destiny, religion, and determinism. The philosophers held speculation sessions for decades!

In hindsight, his principle is not mystical but clear to anyone who studies the properties of waves since the principle is a simple property of a wave train and not dependent on QM. It works as follows:

Suppose we draw two wave trains, A) and B) as in Figure 9-6. The length **Δx** of the train at A) is long and its dominant wavelength λ can be accurately measured because there are lots of nodes to use. The length of the train at B) is short and the few nodes mean that the dominant wavelength λ can be determined only poorly. It is clear that: *if the train size is small, the error of wavelength is large*. This fact can be put into mathematical terms:

error of wavelength = Δλ / λ ≈ 1/(number of nodes), and

the number of nodes = the train length/wavelength = Δx/λ

Put these two equations together and get

$$\Delta x \, \Delta \lambda \approx \lambda^2 \tag{9-4}$$

which says the product of the two errors is a constant, λ^2.

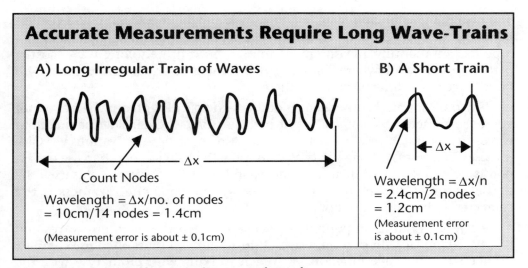

Figure 9-6. Errors of measuring wavelength.
If you wish to accurately measure the wavelength contained in a train of unknown waves, you need to have a large number of nodes to count. The more the nodes, the more accurate is the measuremant. Thus the accuracy of measuring the long wave-train at (A) is about ten times better than at (B). This fact can be used to demonstrate the truth of the Heisenberg Principle $\Delta x \Delta p \approx h$.

This result can be easily translated into the Heisenberg Principle because of two relations: 1) **Δx** is the same as the error of position of a particle in the wave. 2) The wavelength is related to particle momentum through the DeBroglie relation $\lambda = h/p$. We can switch the λ variable into the momentum **p**, by using the calculus of Chapter 2 to find,

$$\Delta \lambda = \Delta p \, \lambda^2 / h$$

Substitute this into Equation (9-4) to get **Δx Δp** ≈ **h**, the Heisenberg Principle. This is what we wanted to prove.

There is another version of the Uncertainty Principle which says that the product of error of time **Δt**, and the error of energy **ΔE**, is also equal to Plank's constant, or

$$\Delta t \, \Delta E \approx h$$

You have probably already guessed that this one can be obtained from Equation (9-4) too, by using the relation **E = hf**.

MATH FOR THOUGHT: Try using E = hf to work out $\Delta t \, \Delta E \approx h$ for fun.

CHAPTER 9/*Quantum Mechanics*

The uncertainty principle states that if you know the momentum of a particle perfectly, then you can have no knowledge at all of the position. This is because a perfect momentum measurement implies an infinitely long wave, so the particle could be anywhere. Similarly, if the energy is exact, you can have no knowledge of the time when it got there.

The philosophical consequences of this principle depend on whether or not you believe there is actually a "point particle" somewhere inside the quantum wave. If you say, "yes," then the conclusion must be that Nature, via QM, imposes a fuzziness on our ability to determine the location of otherwise precise points. If you say "no," then the conclusion is that the wave packet is itself the particle, sometimes smaller, sometimes larger. Then the fuzziness is the character of the particle, not our ability to find it.

There is one final important difference between the quantum waves and the more familiar light or water waves. It turns out you must often use complex numbers for the wave functions Ψ. When working with ordinary waves, the complex numbers make the algebra easier. Then when you are done you can go back to real numbers by taking the real part. But for mysterious reasons, the complex numbers are often *necessary* in QED to get the right answers. No one knows why.

The Ultimate Paradox: Bell's Theorem

In 1935, Einstein, Podolsky, and Rosen (EPR) put forward a *gedanken* ("thinking") experiment whose outcome they thought was certain to show that there existed natural phenomena that quantum theory could not account for. The experiment was based on the concept that two events cannot influence each other if the distance between them is greater than the distance light could travel in the time available. In other words, only *local events* inside the light sphere can influence one another.

Their experimental concept was later used by John Bell (1964) to frame a theorem which showed that either the statistical predictions of quantum theory or the *Principle of Local Events* is incorrect. It did not say which one was false but only that both cannot be true, although it was clear that Einstein expected *The Principle* to be affirmed.

When later experiments (Clauser & Freedman 1972; Aspect, Dalibard, and Roger 1982; and others) confirmed that the quantum theory was correct, the conclusion was startling. The Principle of Local Events failed, forcing us to recognize that the world is *not* the way it appears. What, then, is the real nature of our world?

The important impact of Bell's Theorem and the experiments is that they clearly thrust a formerly only philosophical dilemma of quantum theory into the real world. They show that our ideas about the world are somehow profoundly deficient. No one understands these results and only scant scientific attention has been paid to them. On the other hand, much philosophical discussion has appeared.

The Essence of Bell's Theorem. His theorem relates to the results of an experiment like the one shown in Figure 9-7: A source of two *paired photons*, obtained from the simultaneous decay of two excited atomic states, is at the center. At opposite sides are located two detectors of polarized photons. The polarization filters of each detector can be set parallel to each other, or at some other angle, freely chosen. It is known that polarizations of paired photons are always parallel to each other, but random with respect to their surroundings. So, if the detector filters are set parallel, both photons will be detected simultaneously. If the filters are at right angles, the two photons will never be detected together. The detection pattern for settings at intermediate angles is the subject of the theorem.

Bell (and EPR) assumed that the photons arriving at each detector could have no knowledge of the setting of the other detector. This is because such information would have to travel faster than the speed of light — prohibited by Einstein's Special Relativity. Their assumption reflects the Principle of Local Causes, that is, only events *local* to each detector can affect their behavior.

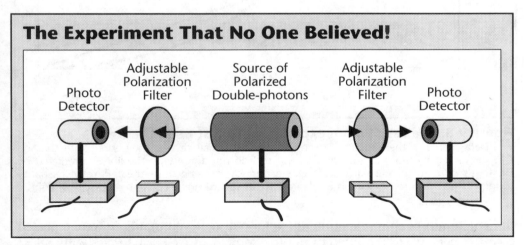

Figure 9-7. Experiment to test Bell's theorem.
Polarized photons are emitted at the center, pass through the adjustable polarization filters on the left and right, and enter detectors on each side. Coincidences (simultaneous detection) are recorded and plotted as a function of the angular difference between the two settings of the polarization filters.

Based on this assumption, Bell deduced that the relationship between the angular difference between detector settings and the detected coincidences of photon pairs was linear, like line L in Figure 9-8. His deduction comes from from the symmetry and independence of the two detectors, as follows: A setting difference of Δ, at one detector has the same effect as a difference Δ, at the other detector. Hence if both are moved Δ, the total angular difference is 2Δ and the total effect is twice as much, which is a linear relationship. The other line labeled **QM** is the calculation obtained from quantum theory.

Bell, EPR, or anyone who does not believe in superluminal speeds, would expect to find line **L**. In fact, the experiments yielded points **R**, which agreed with line **QM**. The predictions of quantum theory had destroyed the assumptions of Einstein, Podolsky and Rosen!

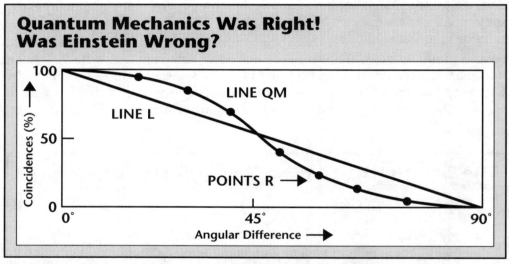

Figure 9-8. The result of an experiment to test Bell's theorem.
Data points of the experiments are shown with small circles. They agree with the line QM, predicted by the quantum mechanics, and do not agree with the line L, predicted by Einstein's concept of casuality. This was a big surprise, because the failure of casuality suggests that the communication is taking place at speeds greater than the velocity of light.

The results of these experiments were so disbelieved that they were repeated by other persons, using different photon sources, as well as particles with *paired spins*. The most recent experiment by Aspect, Dalibard, and Roger, used acousto-optical switches at a frequency of 50 MHz which shifted the settings of the polarizers *during* the flight of the photons, to completely eliminate any possibility of local effects of one detector on the other. Nevertheless, they reported that the EPR assumption was violated by five standard deviations, whereas quantum theory was verified within experimental error (about 2%).

Do Non-local Influences Exist? Bell's Theorem and the experimental results imply that parts of the universe are connected in an intimate way (i.e. not obvious to us) and these connections are fundamental (quantum theory is fundamental). How can we understand them? The problem has been analyzed in depth (Wheeler & Zurek 1983, d'Espagnat 1983, Herbert 1985, Stapp 1982, Bohm & Hiley 1984, Pagels 1982, and others) without resolution. Those authors tend to agree on the following description of the non-local connections:

1. They link events at separate locations without known fields or matter.
2. They do not diminish with distance; a million miles is the same as an inch.
3. They appear to act with speed greater than light.

Clearly, within the framework of science, this is a perplexing phenomenon.

Bizarre Consequences. Some proposed explanations for the result of Bell's Theorem reach out to the bizarre in desperation, as diagrammed in Figure 9-9. One of these explanations, the Many Worlds idea suggests that our world is paralleled by one or more other almost identical worlds, existing in other spaces

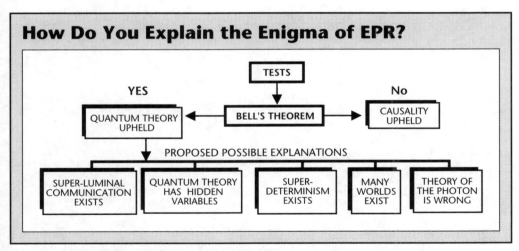

Figure 9-9. The Results of the EPR Experiments.
The last row shows the strange variety of proposals to interpret the surprising results of the experiments which tested Einstein's "EPR" gedanken experiment. Some of these proposals are as paradoxical as the results they intend to explain.
Quantum theory's stronghold as a fundamental law of natural phenomena was upheld. But an equally firm principle from relativity, that no information can travel faster than light, was severely questioned. Scientists are very puzzled by this.

of which we have no knowledge, except for an occasional leak like an EPR event. Another idea, Super-Determinism, is a philosophy which implies that our fates are fixed, and that the outcome of these experiments was already determined by the unknown "Master of Fate." The Super-Luminal theories accept the idea that some things travel faster than light.

Is the Theory of the Photon Wrong? This possibility, suggested in Chapter 13, is that the physics of the photon is incorrect. There is strong evidence elsewhere for this possibility, including the paradoxes of the photon itself, and the dilemma of the point electron which violates conservation of energy. The electron's problems spill onto the photon because it is the thing which carries energy between electrons.

In Chapter 13, the energy-carrying role of the photon is replaced by coupled oscillators, which mimic non-local communication between particles, but do not exceed the speed of light.

"There are more things in heaven and earth, Horatio, than are dreamt of in your philosophy."

– Shakespeare, *Hamlet*

CHAPTER 10

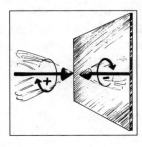

Particles and Electricity

The Simple Properties of the Particles
Discovery and Use of Our Industrial Servant: Electricity

CHAPTER 10
Particles and Electricity

The Simple Properties of the Particles

Without particles to populate it, the universe and its laws could not exist; they would be meaningless. Therefore, before we attempt to understand it, we should also understand the three particles, electron, proton and neutron, which make up the matter of the universe.

One property of anything which we term a "particle" is that of location. We mean that we can identify the position of the thing and be able to say it is "here" and not there. Quantum Mechanics, of course, determines the exactness of location. In this respect, all particles are ruled alike by Quantum Mechanics.

The intrinsic properties possessed by a simple particle are few, typically about half a dozen. To ask about more properties has no meaning. While we might hope for more familiar ways to enhance our mental image of a particle, we must accept that there are no more properties. For example, to ask about the color, shape, or age of a particle is quite meaningless. In addition to these measurable properties, there are others of a generic nature. Of these, parity and class (lepton or baryon) will be discussed below.

The numerical properties of these three particles are listed in Table 10-1.

TABLE 10-1. THE NUMERICAL PROPERTIES OF PARTICLES.

Property	Electron	Proton	Neutron	(multiplying factor)
CHARGE	-1	+1	0	$\times\ 1.6 \times 10^{-19}$ Coulomb
FREQUENCY (MASS)	1	1836	1839	$\times\ 1.237 \times 10^{20}$ Hz
SPIN	1/2	1/2	1/2	$\times\ h/2\pi$ kg-m^2/sec
MAGNETIC MOMENT	1	2.8/1839	-1.9/1841	$\times\ eh/mc^2 = 1.94 \times 10^{-31}$

Some Clues to the Origin of Numerical Properties. Although the origins of these basic numbers are unknown, there are some clues. Hopefully they will lead us to a full understanding of particle origins. Most clues relate to our experience in the human realm, so one should be cautious and flexible in drawing conclusions, especially since we know these particles belong to the quantum realm.

The first clue: Notice how the charge and spin of the electron and proton are exactly the same but that their masses are different. Mass seems to be a special property of each particle, whereas spin and charge are universal. In fact, all of the known particles share this characteristic; charge is always +1 or -1 or 0, spin is +1/2 or -1/2, but masses vary widely. Whatever it is that determines their mass appears not to affect charge or spin.

Another clue: These properties are always factors in the QM wave function of the particles because each of them affects the total energy. Are these properties related to the character of the waves? Is there a "charge part" of the wave which is the same for all the particles, and is there a "mass part," different for each particle? These questions are hard to answer since QM has no physical model of the particles. Further, we don't really understand the waves, even though calculating them is just a matter of following the rules of QM.

The Peculiarity of Spinning Particles. One unique quantum feature of particles is a type of angular momentum called "spin" which is associated with the quantum wave. The values of the spin found in Nature are always multiples of either +1/2 or -1/2 (Remember, "spin 1/2" is shorthand for 1/2 x $h/2\pi$). Thus, identical particles with differing spin are actually different particles, because their waves differ. For example, it is a mistake to say that an electron with spin 1/2 is the same particle as an electron with spin -1/2. The use of the same name is a historical blunder from the days before spin had been measured. When you calculate their QM properties, they are different particles! The error has been built into the language of physics and is hard to avoid, but you should remember that they are different. Spin has units of angular momentum, and if transferred to another particle, the spin can become ordinary angular momentum of the other particle.

A dilemma exists with respect to the rotational character of spin, as follows: Particles are spherically symmetrical in regard to charge, mass, and behavior. In spite of this, having a spin from a human realm view demands a spin axis, which would destroy the spherical symmetry! How can this be? Is there symmetry or isn't there? There might be an escape from this dilemma because whenever spin is transferred in an interaction (i.e spin is measured), the spin axis is always found to be along the line of particle motion. Thus spin might be a result of particle motion, whereas symmetry is a property of the particle.

There is another clue: In the QM mathematics, particles and anti-particles have spins of *opposite* rotation around the line of motion. Perhaps this fact can lead to the idea of the quantum wave structure. But, whatever spin is, it is not the toy-top model which early explanations hoped to find. You might ask, "Why not try to change the spin math?" And the answer, typical of QM, is that you must use a strange spin math because only then do the computations match the measured energy levels of the atoms. Spin is very puzzling and it was this puzzle that led Niels Bohr to the *Copenhagen Doctrine*, described in Chapter One.

Enigma 10-1: *What is the meaning of spin measurements and spin waves?*

In 1928, Paul Dirac devised a relativistic wave equation which yielded two opposite spin values for two electrons; apparently particle and anti-particle. When the positron was discovered, years later, Dirac's Equation was remembered and heralded as prophetic. The equation starts with the relativistic energy equation and yields wave functions with four components like a vector. It is entirely mathematical and gives no hint of any inner machinery which makes the electron work, but it does give precise values of spin and energy in the hydrogen atom.

Spin and the Exclusion Principle. The interference of the waves of particles causes one of the most far-reaching effects in chemistry. The effect was simply stated as a principle by Wolfgang Pauli:

> **EXCLUSION PRINCIPLE:** *No two identical particles with spin 1/2 can exist in the same state together.*

Spin 1/2 particles include the electron, neutron, and the proton. To understand the principle, it must be realized that electron waves are not all alike. Any given electron's wave is distinguished by the values of its frequency (energy), angular momentum, and spin. The principle says if two electrons or protons have the same set of wave numbers, they are in the same state and cannot exist together. No one understands why this principle exists, but it is found to always be obeyed.

This principle determines the organization of the Atomic Table and the properties of all the atoms in it, which together make up all the matter of our Earth and the universe. The manner in which the Atomic Table is built depends on the exclusion principle. As electrons are added around nuclei of increasing charge, each new electron added must have increasingly larger values of energy and angular momentum numbers, to avoid duplicating those electrons already there. There are only two choices for spin number. This requirement is the major factor in determining the varied character and properties of different atoms in the table.

ENIGMA 10-2: *What causes the Exclusion Principle?*

QUESTION FOR THOUGHT: *If you agree that two electrons with different spins are in fact, different particles, how would you reword the Exclusion Principle?*

"Even" Particles will socialize. "Odds" will not. Another feature of spin and interference is that if particles have an even number of 1/2 units so that spin is integral(spin =1,2,3, etc.) there is no exclusion rule. Such particles can exist together in the same state in unlimited amounts. When deciding the spin of a composite particle, you must combine the parts. For example: Helium He^4 is a combination of 2 protons, 2 electrons, and 2 neutrons, each of which have a spin of 1/2. No matter how their spins are combined, the net result is always an integer 1, 2 or 3. He^4 does not obey the exclusion principle. This accounts for much of the fascinating behavior of liquid helium at close to zero temperature.

Magnetic Moment is Nature's Elementary Magnet. The electron's magnetic moment, like spin, is a vector of rotation and are related. The magnetic moment is defined in terms of the human realm property of a magnetic loop; which is an electric current flowing in a circle. The units of a magnetic moment are the (area of the loop) x (current flow). When viewed using our limited perspective of the quantum realm, magnetic moment is strange in several ways. For example, the neutron which is charge-wise neutral, nevertheless has a magnetic moment. This is puzzling by human realm thinking. Where is the moving charge? If we could understand how a wave becomes a human realm angular momentum, the magnetic moment might become clear too.

Another example, the sizes of the proton/neutron magnetic moments in Table 10-1, at first glance seem reasonable. Since the proton wavelength is one thousand eight hundred and thirty nine times smaller than the electron, the equivalent loop area would be smaller. But by human-realm thinking, moment should be proportional to the square of the wavelength, but no, it is only roughly the first power. These facts show that although you often end up in the right ball park, attempts to compare the quantum realm with the human realm usually go wrong.

Let us try to avoid mistakes by following the scientific method initiated by Tycho Brahe. That is, ask, "What is really *measured* when we find a magnetic moment?" Of course, it must be an energy (frequency) exchange, like all measurements. Magnetic moments are obtained from the spectra of radiating atoms when a magnetic field is present, which typically causes a single spectral line to become two lines. The frequency difference of the two lines is the energy caused by the affect of the field on the moment. In the human realm, this would be: Energy = **M B cos** θ, where **B** is the magnetic field, **M** is the moment, and θ is the angle between them. If **M** and **B** are parallel, then $\cos\theta = +1$. If anti-parallel, then $\cos\theta = -1$. Hence the spacing of the two lines is twice the energy. That is all we know. We cannot make human realm assumptions that there is a loop or a current flowing in it! In fact, the concept of a current within an electron is a contradiction of logic, since the electron itself is the smallest element of current according to our definitions and knowledge. All we know is that if the particle is in a magnetic field, there is an energy (frequency) shift. Since we don't know the internal structure of the neutron or proton, we can't say what loops or currents exist at all. We must use this reasoning to avoid errors from unjustified comparisons between the human and quantum realms.

Big is small, and small is big. I think that all of us, as part of our primitive sense of exploration, want to find the size and weight of new objects. Perhaps our primitive emotions are looking for building materials and anticipate using them like bricks. Then as children, we learn rules of experience for size and weight. So, when we arrive in the laboratory we have a prejudice that heavier things are bigger things. This is a common human realm experience, so it is quite a surprise to find in the quantum realm that sizes are just the opposite! Heavy particles are small particles and vice versa. Why does this occur?

In the quantum world the word "particle" as we use it in the human realm is very misleading. It may even be meaningless. The idea appears to be an illogical carry-over from the human realm which occurred in the early days of research. We ought to remove the word from quantum physics, but this is almost impossible since it is embedded in our language where it continues to mislead our thinking. The fact is that quantum particles do not have the property of shape with defined edges nor specific locations. If we seek a useful concept of size, it is the range of the forces which surround them. For example, consider charge or mass forces: these forces fade away with distance r as $1/r^2$, reaching zero at infinity. Thus it can be said that every particle, in some measure, is everywhere in the universe, but *most* of the forces are concentrated close to a center. For a proton or a neutron, their nuclear forces are much more concentrated close to a center. One useful measure of nuclear size is the Compton wavelength, h/mc, of the quantum waves associated with a stationary nucleus. This also applies to the electron. Since m is in the divisor, compton wavelength size is inverse to mass. For a charged electron, another size measure is e^2/mc^2, which is the radius where electric potential energy equals mass energy. Again, size is inverse to mass.

Why does this happen? It is because in the quantum world, you are dealing with waves and frequencies. The wavelength is a measure of size and the frequency is always a measure of mass/energy. These two measures are always inverse to each other according to the fundamental equation of a wave, $\lambda f = c$.

The above examples and equation show that particle size (≈ wavelength) is inversely proportional to mass (= frequency). Thus the relation between mass and size is opposite our human realm experience. So if you remember that waves are the primary characteristic of the quantum world, you usually find reasonable answers.

Everyone hates to give up the particle idea, so they ask if quantum particles have any kind of a core at their center. Experimenters have frequently investigated this by shooting one particle at others and measuring the results. In the case of the electron, bombardment has never shown a central region at all. When one bombards neutrons or protons, one observes only a central region of force between nucleons which is roughly the size of their wavelength, h/mc. That is the character of the nuclear force; strong in a central region and nothing elsewhere. Outside of the central force region there is a $1/r^2$ charge force for the proton but none for the neutron. The nuclear forces are much larger than the electron charge force, as indicated by the energy of formation which is one thousand eight hundred and forty times larger. This is also a quantum affair: If you convert the various energies, masses, and sizes to units of wavelength and frequency, you will notice that they confirm the relation $\lambda f = c$.

What is a Neutron? It has been widely speculated that a neutron could be a proton plus an electron bound together. If this were true it would explain the neutral charge because their two charges add to zero. But notice that neither the masses nor the spins add to the right value. This guess does not work. Although it is tempting to speculate, no one has yet found a structure for the neutron or the proton.

In addition to the properties in Table 10-1, it is known that an isolated neutron will decay into an electron, a proton, and a neutrino with a half-life of about 14 minutes. Such a long life suggests that the neutron is almost stable. Interestingly, if the neutron is joined with a proton, forming $_1H^2$, the two become stable together. Adding another neutron-proton pair to form He^4 results in unusual stability for the whole combination. In fact, all nuclei which are multiples of four: B^8, C^{12}, O^{16}... etc. are especially stable. This is often summarized by saying *Nuclear forces act mostly on their nearest neighbors.*

On the other hand, a large nucleus will decay if there are too many more neutrons than the number of protons. It is as if each neutron needs a close proton partner for maximum stability. The nuclear decay process appears to be the same as for an isolated neutron. The neutron turns into a neutrino, an electron and a proton. But only the electron and the neutrino are ejected, the proton usually remains behind to pair up with another unpaired neutron. The process is called *beta decay* because when Madame Curie first observed it in 1899, she used the name beta particle for the ejected electron.

The protons and neutrons of a nucleus are bound strongly together. To remove one of them from an average nucleus requires an energy of about 10 million electron-volts (MeV), which is called the *binding energy*. Compare this with the small 10 eV needed to remove an electron from a typical atom nucleus and you can appreciate the great strength of the nuclear force. How is this force created? No one knows, and only a few ideas have been suggested. Is it a property of the "space" in which the quantum force waves travel?

RESEARCH GOAL 10-2: *What is the origin of nuclear and other forces?*

Lepton and Baryon numbers are always conserved. Both neutrons and protons are given the generic name of *baryon* which in Greek means "heavy one." At the other extreme, electrons are given the name *lepton* which means "light one." It has been a curious and commonly observed fact that the total number of baryons never seems to change, despite a great variety of nuclear reactions that have been observed. This has led to:

RULE OF CONSERVATION OF BARYON NUMBER:
Protons − antiprotons + neutrons − antineutrons = a constant

Alternatively, if you assign each proton or neutron a baryon number +1, and each antiproton or antineutron is assigned a baryon number -1, then you can just say,

Baryon number is conserved.

A similar rule applies to electrons,

RULE: Electrons − positrons = a constant

Again, you can convert this into a lepton number rule if you wish.

Unstable Particles are Always Sliding Down an Energy Hill. Both leptons and baryons can be temporarily converted into other short-lived particles of greater mass by adding energy and sometimes spin, if needed. These typically have names using Greek letters like *muon, tauon, pion, kaon, lambda, sigma*, etc. There are several dozen of these, and during their lifetimes of micro to micro-microseconds they can undergo a variety of interactions among themselves, changing from one to another.

One interesting fact about the interchanges is that the total mass-energy of the participants always *decreases* at each change. It is just like we know you can slide downhill, but you can never slide uphill. In fact, a little algebra can show that hill-sliding and energy exchanges follow the same natural rule. Many ideas have been suggested as to why energy exchange behaves this way. Some of the arguments are very plausible, but none are a convincing certainty.

Another interesting fact about leptons and baryons is that no matter how complicated an exchange process is undergone, the lepton or baryon number is conserved throughout the process, so that they finally decay back to a stable baryon or lepton at the end.

Sometimes, if both a baryon and an anti-baryon are involved, the process ends up with both of them annihilating each other leaving only the equivalent mass-energy which is transferred to other particles. Leptons can do the same thing. In an inverse process, a particle and its anti-particle can be created out of only energy, if it is available. The baryon and lepton numbers are conserved since they are equal and opposite in the anti-particle.

What does this pattern of conservation mean? It is very tempting to conclude that there is something especially basic about the structure of baryons and leptons because Nature prefers to keep them intact. Nevertheless, no one knows of such a structure or why the conservation rules exist.

ENIGMA 10-3. *What are baryon and lepton numbers and why are they conserved?*

Discovery and Use of Our Industrial Servant: Electricity

The experimental facts of electricity were discovered in the nineteenth century and framed in mathematical form before the year 1900. The pioneers of classical electricity were Charles Agustin Coulomb, Michael Faraday, Hans Christian Oersted, and Andre Ampere. Each of them deduced separate laws which bear their names. The final step was taken by James Clerk Maxwell (1831-1879) a Scottish physicist who combined their work into four classical laws of electricity and magnetism. These laws have stood since his time and today form the basis of all electrical engineering. They are human realm laws.

It is frequently asserted that Maxwell's Equations are part of the basic laws of the Universe. This is not so since it can be shown (in the Appendix) that they are derivable from the Coulomb law of electric charge plus the rules of Special Relativity. Accordingly it is not necessary to put them on the list of basic laws. Nevertheless, because electric phenomena is so closely connected with the basic behavior of the universe, these four equations are described and illustrated in Figures 10-1 to 10-3.

How Maxwell put Electricity into a Nutshell. The first law, $\nabla \cdot E = q$, shown in Figure 10-1 simply states that the total electric field which emanates from a charge is proportional to the charge itself. Since the E field has this

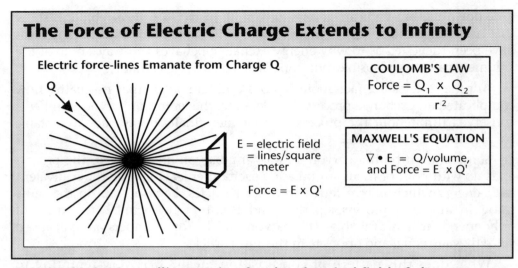

FIGURE 10-1. Maxwell's equation for the electrical field of charge.
The most fundamental law of electricity is also the simplest. It is Coulomb's Law for the force between particles shown in the box at the right. The same law is restated in a different way by defining an electrical "field" E, due to charge Q, and saying that the force on another charge Q' is the field E multiplied by the charge Q'.

Maxwell's Equation offers yet another way to express it, using the elegant mathematics of vector algebra. There is no new law, just a different method. All the rules of electricity and magnetism can be obtained from only Coulomb's Law by combining it with the law of relativity.

property at any distance from the charge the law is also equivalent to saying that the imaginary electric field lines coming from the charge will never end — an easy memory device. It is just a more elegant version of Coulomb's force law.

The second law describes magnetic fields **B**, and is written $\nabla \cdot \mathbf{B}$ analogous to the similar electric field law, except that the vector expression equals zero because there is no magnetic equivalent of a charge. The easy way to understand this law is to imagine that the lines of magnetic field **B** never end but close on themselves, and always arrange themselves as smoothly as possible.

The third law of Figure 10-2 contains the principle of electric generators and motors; a changing magnetic field induces a circular electric field to appear in the space around the changing field. Diverse schemes to change the field such as moving magnets, changing currents, and rotating coils result in a variety of powerful electric machines, transformers, and relays, as well as the gentler speakers, turntables, pickups and tapes of hi-fi music. It can be correctly said that this law together with heat engines are the backbone of the industrial revolution and our modern society. They provide the enormous power of machines replacing puny human muscle power.

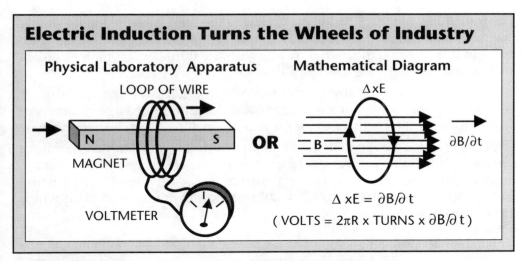

FIGURE 10-2. Maxwell's equation of electric induction.
On the left is a familiar example of an electric field which is induced in a loop of wire by thrusting a magnet into the loop. On the right is shown the diagram of the Maxwell equation which describes this induction. The Maxwell equation states that whenever a magnetic field B changes its strength, at a time rate $\partial B/\partial t$, there is a circular electric field E created in the surrounding space, perpendicular to the magnetic field, and proportional to the rate of change.

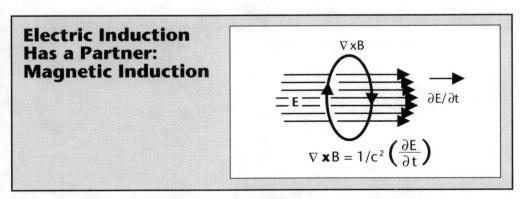

FIGURE 10-3. Maxwell's equation of magnetic induction.
Magnetic induction corresponds exactly to the electric induction except that the roles of the two fields are exchanged and the effect is diminished by the factor $1/c^2$. There is no convenient laboratory illustration of this, so no apparatus is shown.

Stated in words, an electric field E which is changing at a rate $\partial E/\partial t$, creates a circular magnetic field B, perpendicular to the E field, in the space surrounding it, proportional to the time rate of change E.

CHAPTER 10/*Particles and Electricity*

The fourth law of Figure 10-3 is the inverse of the third law; a changing electric field induces a circular magnetic field around it. Industrial applications of it are rare, but it is an equal theoretical partner with the third law. I couldn't think of any apparatus to illustrate the figure. Can you?

Maxwell's Equations always give the correct results in any human realm laboratory experiment. But since they are a description of large numbers of quantum events, it is a mistake to use them in the quantum realm without caution.

This concludes our discussion of the basic laws of Nature. We have determined that all the branches of the tree of knowledge (electricity, biology, magnetism, mechanics, optics, chemistry, etc.), can be obtained using appropriate mathematics with only with the following fundamental laws which lie at the roots of the tree:

1. Newton's 2nd law: Force = d(momentum)/dt
2. The Coulomb law of electric charge force,
3. Newton's law of gravitational force,
4. The rules of Quantum Mechanics,
5. The rules of Special relativity,
6. The conservation of energy, and,

the properties of the fundamental particles: electrons, neutrons, and protons.

Today, science does not yet understand the origins or causes of these laws, nor know of any "inner machinery" which makes them work. The next chapters begin to look into the future and the basis of these laws.

PART II: The Future Land of the Explorers
Chapters 11 to 14

The preceding chapters have described the fundamental laws of Nature as believed to be true by most scientists. The emphasis has been to carefully understand the foundations of knowledge so that the reader can explore further and determine their causes and origins.

At this point in our search, which has already pointed towards the fabric of space as holding the key to the enigmas we have found, we must begin to look into the future.

Future Adventures Exploring the Physics of the Universe

The future is only to be seen in a cloudy crystal ball. Measurements of objects at distances beyond our own Solar System are difficult to obtain. As a result our knowledge is limited and the conclusions that have been assembled are tentative. You must exercise judgement and constraint when forming opinions.

In the next chapters I will describe the probable land of the future explorers and current ideas about it. These chapters discuss:

The universe, its size, shape, and geometry and the particles which make up its matter and give it properties which we can observe. This is a discipline usually termed cosmology.

Possible Conservation Laws of the Universe. Are the conservation laws we see in the laboratory true for the whole universe?

Unresolved proposals of relations between the Universe and particles. One hundred years ago, Ernst Mach and others and 60 years later, Paul Dirac, began to see connections between the very large and the very small, and made intriguing proposals which have remained unresolved in spite of many attempts.

The Big-Bang Theory. It is claimed that the entire universe began in one gigantic explosion in a split microsecond, thirteen billion years ago. Not only that, it happened in a volume of almost zero size! Is this believable?

A Space-Resonance Theory. These are my ideas on how to explain the machinery behind the basic laws of physics. It turns out that these ideas also explain some of the puzzles of cosmology.

Who is the Competition?

At this time, scientists are busy in many areas related to important human needs in medicine, food, housing, and communication. Only a few persons are working in fields related to goals of this book, which are fascinating and excite our curiosity but offer no prospect of immediate return on invested capital. Some related fields have attracted a lot of attention even though their investment prospects are also dim. These are:

High-energy physics. This experimental branch of the tree of knowledge begins with the neutron and proton, seeking their structure and that of their heavy cousins. Big accelerators create transient high-energy states of the proton, which are compared with a structure theory called the "Standard Model." The theory models the heavy particles, termed *baryons*, as made of lesser particles called quarks. Some of the well-known contributors to this model are Richard Feynman, Steven Weinberg, Murray Gell-Mann, T. D. Lee, and C. N. Yang. Paul Davies has written several serious and enjoyable books.

General relativity and Gravity. Einstein's general relativity and some of its mathematical solutions are the heart of a family of cosmologies which explain the structure of the universe. Gravitation is the assumed cause of the possible shapes of space. Some popular consequences of the theory are the familiar black holes, gravity waves, and doorways to other worlds. These and other results remain in the limbo of proposed theories since none have been observed. A famous author of very readable books, a student of the black holes, is an Englishman Stephen Hawking.

NASA's Hubble Space Telescope. A major hindrance to learning more about space is the optical barrier of our atmosphere through which we can see only with light restricted to a narrow wavelength band. This NASA telescope, avoids that problem. Be prepared for new measurements and exciting information from it.

Scientists of the future 100 years hence, may chuckle quietly when they read today's naive theories. Progress is that way. But since intelligent curious explorers are reading this book making plans for the year 2000+, that progress has already begun.

Cosmology is the most perfect science, because there are so few data to disprove it!

— Astronomer's witticism

CHAPTER 11

The Universe

The Most Perfect Science
Have We Found the Size of the Universe?
Geometries of the Universe Are the Paths of Light Rays
Are There Conservation Rules of the Universe?
Mach's Principle Implies Newton's Law of Inertia
The Mysterious Large Numbers of Paul Dirac
The Amazing Theory of a Big-Bang That Began the Universe

CHAPTER 11
The Universe

The Most Perfect Science

Modern scientific cosmology is one of our grandest intellectual adventures. It grasps all the elegance of physics applied on the largest possible scale. Cosmology deals with the structure and mysteries of the universe. The few data we have tantalizingly suggest that there is an intimate connection between the properties of the universe, the largest physical thing that we can conceive, and the properties of the fundamental particles, the smallest physical things we know. So, the universe is very important to the study of fundamental science and the goals of this book.

It is so big, and as yet so unexplored, that we are on uncertain ground when we try to make scientific deductions from the meager knowledge that we have. As the tongue-in-cheek witticism beginning this chapter indicates, there is lots of room for elegant theories constructed from beautiful concepts, or mathematical constructions or geometric comparisons, which rest on such tiny scraps of evidence that they can logically be neither proven nor disproven. The only way out of this dilemma is to patiently await the gathering of further evidence or to make better use of the evidence we have.

Cosmology employs the physical laws established on the Earth and within the Solar System. This may not be correct. We have no logical reason to assume that these same laws apply at larger distances and this skepticism should increase in proportion to the distance from us. Since the Solar System is so unimaginably tiny compared to the vast reaches of space, it should not be surprising if someday we find the physical laws are different out there. In the meantime, we have no other choice but to make the assumption that they are the same as on Earth, or similar.

A complete treatment of cosmology should include Einstein's general theory of relativity which for computation and analysis uses the elegant but advanced tensor and vector calculus. This elaborate mathematics is beyond the goals of this book. Fortunately, the basic principles and results, which describe the behavior of matter and light under the influence of gravity, can be comfortably discussed with ordinary algebra. Einstein's theory implicitly assumes that the structure of space (the shape of the path of a light ray) is determined by the gravity fields of the matter presently in space. This idea is easily grasped since the amount of "space-bending" is proportional to the amount of matter present.

It should be kept in mind that the results of general relativity have only been partially verified by measurements. Even these measurements can be explained by special relativity without the general theory. In contrast, special relativity has been amply verified for the theory's prediction of the increase of energy and mass depending on relative velocity, *but not yet for the dilation of time or increase of length, discussed in Chapter 4.*

Have we Found the Size of the Universe?

Before this century, there was almost no astronomical knowledge of space beyond our own solar system. Astronomers were aware of the milky way and it was correctly conjectured that the stars in it formed a disk-like shape, our galaxy. Most thought that our galaxy composed the entire universe, which was not an unreasonable idea since the number of stars contained is about a hundred thousand million. Quite a lot. A galaxy similar to ours is shown in Figure 11-1, where you can imagine our Sun as an average yellow star near the inner side of one of the rotating spiral arms.

FIGURE 11-1. A Typical Spiral Galaxy of Stars.
A typical galaxy contains about 100 billion stars, and there are estimated to be about 100 billion galaxies in the whole universe. From this information and knowing the mass of a star like our sun, you can estimate the mass of the universe.

If this galaxy were our own, the Sun would be about halfway out from the center, in a spiral arm. From our earth we can look out and see the other stars of our galaxy forming a band in the night sky, which we call the milky way.

Try to imagine how large, or small depending on your view point, you and I are, living on this tiny Earth in this fantastically enormous universe.

Edwin Hubble was the Pioneer of Galaxies. Our modern concept of the universe dates to 1924 when an American, Edwin Hubble, started to find out if ours was the only galaxy. He knew that there were many objects in the sky which had galaxy-like shapes. In order to prove they were galaxies, he had to determine their distances. He noted that certain types of stars which are close enough to measure always have the same luminosity. So he argued that such stars, if in another galaxy, would also have the same luminosity, and by measuring their brightness he could calculate their distance. This assumption could be checked by measuring several stars in the same galaxy which must yield the same distance.

Hubble worked out the distance to nine different galaxies, which influenced many other observatories to also measure galaxies. Stars are so far away that they appear just as pin-points of light. The only way we can distinguish one from another is by the color-spectra, intensity and polarization of the light. Astronomers measured all these properties and found to their surprise that the spectral lines were nearly all shifted by varying amounts towards longer red wavelengths. Such a shift is expected if there is a speed difference between the emitter and the observer. This effect, known as the Doppler shift, is an everyday experience in the case of the tonal shift of sound of a train whistle. But everyone expected random variations among the stars and as many blue shifts as red.

Hubble's Revolutionary Proposal. Hubble spent five more years cataloging the spectra and distances of galaxies, and in 1929 announced that the red shift was proportional to the distance from us. This was quickly interpreted to mean that the universe was not static as most persons had previously thought, but that the universe is continually expanding. He proposed a relationship $v = Hd$, where d is the distance from Earth, and H is a constant, now called *Hubble's constant*. Using the most recent data, the value of $1/H$ is $\approx 1.3 \times 10^{10}$ years. Assuming that the measured Doppler shift is an actual measure of velocity, this relation suggests that all objects are receding away from each other, with a velocity v that increases with distance between them.

This discovery was a great milestone of 20th century astronomy. Only 60 years ago, telescopes had not identified any objects outside of our own galaxy. Our galaxy was thought to be the extent of the entire universe, so no one knew of the enormous distances beyond. And even today, its actual size, if the concept of its size can be imagined, is only conjectured from the derived measurement of the *Hubble Distance*.

The above conclusion assumes that Earth is not at the center of the universe. Rather than assume that the Earth is at the center, an unlikely remarkable coincidence, it is more modest and logical to assume that all of space is undergoing a continual expansion. Thus, no matter where you measure from, you should always obtain the same measurement of expansion as did Hubble. This interpretation is popular among most scientists.

Varied Interpretations of the Hubble Measurements. If you further assume that this expansion has been present since the beginning of the universe, then the farthest objects from us are at a distance that light could travel in that time, or 1.3×10^{10} light-years. This distance is: $d = ct = 3 \times 10^8$ m/s $\times 1.3 \times 10^{10}$ yr $= 3 \times 1.3 \times 365 \times 60 \times 60 \times 24 \times 10^8 = 123 \times 10^{24}$ meters. This is a common estimate of the radius of the universe. Reasoning in the same way, one can also speculate that the age of the universe is $1/H$ or 1.3×10^{10} years.

There are other interpretations of the observed spectral "red shift" of light from distant objects. It might be due to some gravitational effect. That is, the light has passed through a gravitational field and in so doing has lost energy ΔE, which shows up as a decrease of frequency, according to $\Delta E = hf$. Such fields have not been observed. Another possibility is that space itself varies with distance from us, so that the speed of light c is not constant, and the basic relation $c = f\lambda$ depends on distance from us. Many other suggestions have been made.

Yet another possibility is the proposal I make in Chapter 13 which depends on an assumption that *gravity is repulsive between matter and anti-matter*. This allows a universe in which matter is created gradually starting from an empty universe. Physical constants of the atoms increase in proportion to the total matter present, so that the energy exchanges between atoms begin with near zero frequency at the beginning of matter formation and increase as time goes on. This theory then interprets the red-shift as a result of observing distant stars which radiated at an earlier time with longer wavelengths.

There is no present way of making direct measurements of galactic distances. Methods of indirect distance measurement use super novas, sizes of galaxies, and "Seyfert" clusters, but these depend on assumptions which are not certain. So none of these possibilities have been proved or disproved.

It is an Awful Lot of Matter! There are clever guesses about the matter content of the universe. About 10^{11} galaxies can be seen by the best telescopes and each galaxy contains about 10^{11} stars, and each star contains about 10^{57} hydrogen atoms. If we speculate that the total number of galaxies is actually about 10^{12}, we find the total number of protons in the universe is about 10^{80}. Even this guess has to be tempered by the fact we are unable to see matter which does not radiate light, which significantly adds to the total.

In the years since their discovery galaxies have become among the the most vigorously studied astronomical entities. Oort, in 1932, discovered that most of a galaxy is invisible. A huge outer halo that emits no visible light surrounds the visible portion of spiral galaxies like ours. This halo contains most of the gravitational effect of the galaxy and thus apparently the matter. No one has come up with a certain explanation of what kind of matter is in this halo. This is important since it may make up most of the matter of the universe.

Galaxies Have Strange Shapes and Structure. Some galaxies stubbornly refuse to fit into standard shape categories. But some of the unusual shapes can be elegantly explained by computer simulations of catastrophic galaxy encounters. Two close galaxies NGC 4038 and 4039 have long tails and nudging heads like enormous flagellated bacteria. The computer patterns produced by Alar and Juri Toomre at MIT and the U of Colorado neatly reproduce those shapes. So, it was concluded that the two galaxies are approaching each other in elliptic orbits with gravitational interaction creating the huge curved tails.

Other galaxies appear to be undergoing violent internal activity with a seething nucleus at their center. The centers are intensely bright with low frequency radio emission. Similar galaxies also emit immense focused streams of ionized gas that protrude from the nucleus. An example is NGC 5128 in the constellation Centaurus A, which emits over a *million* times as much energy as a typical galaxy. Explanations have been proposed which assume black holes at the center and magnetic fields to focus outgoing streams of ions.

Are There Galaxies of Galaxies? Galaxies are not arranged randomly but tend to exist in groups and clusters; a surprising result of recent sky surveys. At least four superclusters, organized structures composed of many clusters of galaxies, have been found. Each cluster may consist of hundreds to thousands of individual galaxies. For example, photographs made by Laird Thompson at Palomar Observatory show that the galaxy NGC 4535 in the constellation Virgo is a member of a supercluster of which our own galaxy is probably a member. Strikingly, at the other extreme, the surveys reveal that the huge volumes of space between the superclusters are quite empty.

A word of caution: many of the deductions concerning galaxies have assumed that the red shift of their light spectrum, using the Hubble constant, correctly determines the range of objects studied. This is the best judgement available but is not certain.

A Quasar is a Fantastic Source of Energy. Since the 1969 discovery of the now legendary quasar 3C273 by Maarten Schmidt, they have remained an outstanding enigma of astronomy. Some quasars have a red shift of $3\frac{1}{2}$ times the wavelength of the normal spectral light emitted. If one interprets this red shift as due to a receding Hubble velocity, then their distance from us is about 17 billion light years, placing them near the far edge of the universe. A calculation of the total energy emitted yields the fantastic result that they have the intensity of 10^{50} of our Suns. Another remarkable property of the quasars is that they occur about 1000 times more densely in far space (distances equivalent to a red shift of 2X) than they do in our own vicinity.

These astounding results have led to controversy among some astronomers over whether the expanding universe interpretation of the Hubble measurement is correct. Some maintain that unexplained physics is at work and that the quasars are much closer than the red shifts indicates. Theories involving the continual generation of new matter have evolved. Other astronomers have continued to search the region of the quasars, found other galaxies, and conclude that the quasars are indeed supermassive objects at the center of super-galaxies.

Geometries of the Universe Are the Paths of Light Rays

One of the goals of galaxy-searching is to measure the density of matter in the universe and inject that datum into general relativity to find the geometry of the universe. Since the universe appears to us as mostly empty space, it is difficult to imagine at first what is meant by its geometry. Imagination can be helped by thinking of the paths followed by light rays which trace out the grid lines of the space in a geometry. Three frequently discussed geometries are mathematical results of general relativity which assumes that light paths are determined by the matter density in space.

An easy way to visualize the geometry of 3D-space is to first think about the 2D geometry of a *plane* (a flat sheet), then think of the *curved* surface of a sphere, and finally a *curved* saddle shape. Imagine light rays traveling so that they follow the two-dimensional surfaces. Then you can deduce that two geometric properties of a flat plane are that its flat surface is unbounded and its area is infinite. The surface of a sphere is quite different; its curved surface is bounded, and its area is finite. The saddle shape on the other hand is curved, unbounded, and the area is infinite.

Now think about rays in 3D space. According to general relativity, there are also three possibilities for the geometry of the space of the universe, analogous to 2D space. They are *flat* space, *closed* curved space, and *open* curved space. According to the theory, the geometry is determined by the amount and distribution of matter (mass) within it. Where matter is most concentrated, space is most curved.

If space is *closed*, the gravitational attraction among all the galaxies is strong enough so that the Hubble expansion will eventually be stopped and reversed to become a contraction. *Closed* space also curves back upon itself, so that a light ray, after a long enough journey, would return to its starting point from the other direction.

If space is *flat,* the expansion continues, but becomes progressively slower so that in infinite time, it just comes to a halt. The force of gravity on the expanding matter is just strong enough to stop the velocity of expansion at the end of the infinite journey. These light rays travel in straight lines away from their starting point.

If space is *open,* the force of gravity is too weak to ever halt the expansion, which continues forever. The space is curved like a saddle or the edge of a scallop shell, and light rays go on forever curved, but never returning.

A comparison of these three cases may also be made with a launched rocket, which has initial kinetic energy either *lower, equal to, or larger* than the escape energy. If its energy is lower than the escape energy, the rocket will eventually go into a bounded satellite orbit around the Earth (it is like closed space). If its energy is equal to the escape energy, the rocket will come to a halt just as it "escapes" or leaves the last little bit of gravity of the Earth (like flat space). If its energy is greater, the rocket will slow down but never stop, continuing on through space forever (like open space).

Many other geometries can also be imagined depending on matter density. For example, a rough and bumpy space is possible. However, the mathematics are too complicated so these three choices occupy the most attention.

The Type of Geometry Depends on the Critical Density. In general relativity, the Hubble expansion constant is affected by the mutual gravitational attraction of the matter present, and therefore involves the curvature of the space. This is expressed by a relationship between a critical density of matter ρ_c, Hubble's constant **H**, and Newton's constant of gravitation **G**:

$$\rho_c = 3H^2/8\pi G$$

Using recent data, ρ_c is approximately 8×10^{-30} g/cc, a number with accuracy no better than the value of **H** which went into it. It is this critical density of the universe which distinguishes the three types of geometry. If the actual density is critical, space is flat. The actual measured density of space, including estimated galaxies, and interstellar hydrogen, is not far from this value, being about 10% of it. There may also be other unknown sources of matter. It is interesting that the estimated density is so close to the critical density. It suggests that there is some actual meaning or connection between space, the matter in it, and the Hubble measurements.

Are There Conservation Rules of the Universe?

We would like to know if the conservation rules for matter here on Earth also apply to the entire universe. Remember that the Earth-bound rules apply to a definite amount of matter free of external forces, termed a closed system. In this closed system, energy, momentum, charge, etc. are always constant. Is this also true for the universe? Does it behave like a closed system? Since there is not yet much evidence opposing this idea physicists often make the assumption that the universe is a conserved closed system. But this is not a certain conclusion.

The Zero Sum Idea. We have found here on Earth that the values of some conserved quantities add up to total zero. For example, the total charge of all ordinary objects like books, houses, etc. is zero — the objects are electrically neutral. This is because the + charge of each proton in a hydrogen atom is balanced by the − charge of the electron around it, and because the rule of pair production requires that each electron created is accompanied by a positron. All charges sum to zero. Similarly, we ask, is there a zero sum rule for other properties of the universe?

Zero Sums of Momentum and Charge. To say with certainty that the total momentum and charge of the universe is zero is beyond our capabilities, but it is consistent with existing measurements of radiation, spectra, and positions of celestial objects. In localized regions, charge or momentum may become non-neutral, but we can always identify a balancing charge or momentum change of the opposite sign, somewhere nearby, that maintains overall neutrality. So tentatively, it is possible to presume their sum is zero.

A Zero Sum of Matter, too? Edward Tryon (1973) proposed that the sum of energy and matter is also zero in the universe. At first, this seems ridiculous when we view the large amount of matter of all the stars, but analysis yields another result. His argument is that the total amount of matter (a positive quantity) is exactly balanced by gravitational energy between the stars (a negative quantity). He deduces that this would be true if the matter density of the universe is exactly the critical value, as follows:

The energy of gravitation of a particle of mass **m**, acted upon by the rest of the mass M_u, of the universe, from a distance **r**, is

$E_g = -m\, M_u\, G/r$

If we set the mass of the universe equal to the mass contained within a sphere of Hubble radius, **$R = c/H$**, choose mass density to be critical, and set the average distance to the mass to be half the Hubble distance, or **$r = 1/2\, c/H$**, then we get,

$M_u = 4/3\, \pi\, R^3 \times 3H^2/8\pi G = c^3/2GH$,

and inserting these into the expression for the gravitational energy, then,

$E_g = -\, mc^2$

This is a bit amazing! We see that the gravitational energy of a mass particle is just equal to its mass energy, as proposed. The reason the gravitational energy becomes so large is because the universe is very large.

Try Another Cosmology Game. One of the reasons cosmology is so popular is that you can have fun saying, "What if...?" Let's push Tryon's idea further. Think about what happens if all the matter is moved an infinite distance apart. Then the gravitational force between them becomes zero and their gravitational energy rises to zero. Where has the energy input gone? It has gone to moving the matter, so that now the total energy is zero and each mass is zero, m = 0. This geometric example suggests that the geometric meaning of infinity is a distance so large that one matter particle cannot affect another.

Suppose the universe was very tiny and only a few kilograms of hydrogen existed in the universe. Now, the conservation of energy requires that the mass of the H atoms be very tiny to match the tiny value of gravity. We have to conclude that the *constant of gravity is determined by the mass of all the atoms existing in the universe*. We have arrived at Ernst Mach's Principle by logical argument along a different path!

Here is another question involving the tiny universe: What would happen to the electric forces between the protons and electrons? Since no one knows the origin of these forces, you cannot be sure. In later chapters, I propose that they would decrease because they are also proportional to the total matter present. The conclusions obtained explain many of our enigmas.

The preceding analysis is not rigorous, since the net energy of the universe cannot be precisely measured at present, nor is the measurement of Hubble's constant certain or precise. Nevertheless, the argument strongly suggests: 1) the universe is closed. 2) A relation exists between forces and the total mass energy of matter. And 3) $E_g = -mc^2$.

Is There a Zero Sum of Baryon and Lepton number? In order that the sums for baryons (mostly neutrons and protons), and leptons (mostly electrons) may equal zero, we must assert that for every proton in the universe, there exists an anti-proton and for every electron there exists a positron. A proton has baryon number +1 and an antiproton has number -1. Similarly for electrons and positrons.

In our solar system, neither the baryon nor lepton number sum is zero because the Sun and planets are made only of matter. No anti-matter is found. Likewise, the analysis of particles captured in space suggests that our entire galaxy is also only matter. Further, we know that anti-matter and matter cannot exist together since they would annihilate each other. Nevertheless we observe that when matter is created from energy, an equal amount of anti-matter is always created. So we are puzzled as to the location of the anti-matter that should have been created together with the matter of our solar system and galaxy. The only possibility seems to be that some galaxies are made of matter and some of anti-matter. We know that the light from both kinds of galaxy is identical, so we cannot tell the difference using telescopes. Searches by instrumented spacecraft for anti-particles from other galaxies have been fruitless, but that is expected since anti-matter would be ejected from them only by rare explosions.

What Happens if Anti-gravity is Repulsive? Nothing is known about the gravity of anti-matter. I propose, in this book, that matter and antimatter are in fact *gravitationally repulsive*. This makes it possible for equal amounts of matter and anti-matter to exist together in the universe and accounts for only matter being found in our galaxy. According to this idea, when pairs of particles are created, they both form electrically neutral atoms. The anti-atoms and atoms repel each other gravitationally and separate. The anti-atoms leave our Solar system and galaxy and move towards another galaxy composed of antimatter, since it is gravitationally attractive.

There have been no measurements whether gravity is + or − for antimatter, but several experiments have been proposed, and in 1989 through 1991, the University of New Mexico will attempt the measurement. The results may be exciting for cosmology. If it turns out to be repulsive as I propose, our concept of the universe will be drastically changed. We will have to consider mutual repulsion of galaxies and anti-galaxies as a cause of the Hubble expansion. This will compete with the present assumption of initial momentum from the Big-Bang theory.

Another consequent change of our ideas would be that the average curvature of the universe is zero, since for each galaxy there is an anti-galaxy with anti-gravity. However, we cannot be sure of this, since we also have no measurements of the affect of anti-gravity fields upon the path of a light-ray. This concept is discussed further in the section on the origin of forces in Chapter 13.

Another very popular theory suggests that indeed the universe is mostly matter. One of these is the Big-Bang concept, discussed at the end of the chapter, in which it is argued that high-density conditions in the first 10^{-35} second of the bang led to an asymmetry of matter which has prevailed ever since.

Mach's Principle Implies Newton's Law of Inertia

In 1893 Ernst Mach studied the rotation of the Sun, planets, Moon and stars and came up with the startling conclusion that Newton's law of inertia depends upon the existence of the distant stars. No one has yet proven the idea he put forth but every cosmologist has thought about it and taken a position. Einstein accepted his idea and thought that he had incorporated it into general relativity. Here is the basis of Mach's idea:

There are two fundamentally different methods of measuring the speed of rotation of the Earth, or any rotating object. First, without looking at the sky, one can use a gyroscope to measure rotation relative to the inertial space in which the gyroscope spins. The gyroscope essentially finds the centrifugal force on a mass m due to rotation and uses Newton's Law in the form $F = mv^2/r$ to find circumferential rotation speed v. It is similar to the well-known centrifugal force we feel when spinning on a merry-go-round. We do not understand the origin of the law or the related property of space, but frequent use has unquestionably shown the method to be correct.

The second method of measuring the Earth's angular rotation is to simply compare the Earth's position with the "fixed" (distant) stars. It is well known that both of these independent methods give exactly the same result. Is this a remarkable fact which demands investigation? Mach and Einstein thought so. Or is it simply a coincidence?

Opinions differ widely on this question. Some feel that the coincidence is a result of Newton's laws or relativity and of no special interest. Einstein did not think so. He thought that it was of fundamental importance and any comprehensive explanation of inertia or cosmology must encompass it.

Mach reasoned that there must be a causal connection between the existence of all the distant matter in the universe and any local inertial reference frame. Therefore, he assumed what is now known as Mach's Principle: *Every local inertial frame is determined by the composite matter of the universe.* The principle implies that the magnitude and direction of inertial forces on a local mass is determined by relative motion with respect to the total mass of the universe.

Can anything further be deduced about the possible connection? Yes. Since the measure of inertia is the mass of a body, it seems probable that the effects of the stars should be proportional to their mass. Einstein (1917) wrote in this connection, "In a consequential theory of relativity there can be no inertia of matter against space but only matter against matter. If a body is removed sufficiently far from all other masses its inertia must be reduced to zero." He presumed that the effect diminishes with distance for each mass.

Experiments do show that the effect of the Earth's mass on local inertia is too small to be measured, which agrees with a calculation showing that the Earth's mass is negligible compared with that of the known universe. The fact that inertia is the same in all directions also suggests that the universe is isotropic on a large scale; thought to be true for other reasons. In fact, the observed large scale isotropy is termed *The Cosmological Principle*.

Let's try to extend Mach's Principle to other forces. Assume the Principle is true, then distant matter must determine the gravitational mass of a body too, because painstaking measurements have shown that inertially measured mass using $F = ma$, and gravitational mass, using GMm/r^2, are exactly equivalent. Notice that this is the same conclusion arrived at independently above in the section discussing a zero sum of mass.

Is it possible that electric forces are similarly determined? It is logically reasonable since both gravitation and electric forces have the same dependence on distance, $1/r^2$, suggestive of a relation with distant matter.

These conclusions lead to another paradox of mass. Most persons feel, and textbooks write, that gravitational mass is an intrinsic property of an object. On the other hand, inertial mass is usually explained as a property of both the object and the external inertial frame of reference. But if both kinds of mass are equivalent, mass cannot be both extrinsic and intrinsic. Which is it? Or are both wrong, and Mach right?

Paradox 11-1. *What determines the property of mass? Space? Mach's Principle? Or something else?*

There may be an experimental way to measure an effect of Mach's Principle, due to the elliptic shape of our galaxy. If Mach's Principle is true, the magnitude of inertia would not be the same in all directions relative to the galactic plane. Rough calculations show that the uneven mass of our elliptical galaxy could cause a difference of size roughly 10^{-7} between inertial vectors measured perpendicular and parallel to the galactic plane. The experimental precision required lies just on the edge of present measurement technique. Recent advances at the Inertial Guidance Laboratory of MIT might make this valuable experiment possible.

Research Goal 11-1. *Measure the Effect of our Galaxy on an Inertia Vector.*

Dirac was a professor of physics at Cambridge University, England occupying the same honorary chair held by Isaac Newton 250 years earlier. Like Newton, and the readers of this book, he was intensely curious about the diverse facets of Nature and sought to understand the origins and causes of facts he observed. One of these was the origins of the natural constants: c, e, m, m_p, G, and H. He noticed that different ratios of the constants formed very large dimensionless numbers and surprisingly had values close to each other. In 1937, Dirac proposed that these were not coincidental, but suggestive of a mysterious relation between microphysics and cosmology. Two of these ratios are:

The Mysterious Large Numbers of Paul Dirac

$$e^2/Gm^2 \approx 10^{40} \approx cT/(e^2/mc^2) \text{ where } T = H^{-1}$$

At left is the ratio of charge force and gravity force for electrons. On the right is the ratio of the radius of the universe and the electron. Both ratios are approximately equal to the very large number 10^{40}. No explanation has been found for this unusual coincidence, but many persons have tried to unravel the mystery.

Dirac also argued that his hypothesis led to the conclusion that the constant of gravity **G**, must be changing with time. He assumed that **e**, **m**, and **c** were fixed, so therefore **G** depends directly on the Hubble constant **H**, which changes in most cosmological models.

Dirac found another mysterious large number relation:

$$G\rho_u / H^2 \approx 1 \text{ where } \rho_u = \text{density of the universe.}$$

Why should the ratio of these enormous numbers be the unlikely value of one? Chapter 13 makes a suggestion.

The Amazing Theory of a Big-Bang that Began the Universe

Following Hubble's measurement of the red shift of light from distant stars, and his conclusion that the universe was apparently expanding, many persons have asked themselves when and how the expansion began. The immediate conclusion using the Hubble constant, was that the universe began 1.3×10^{10} years ago in a very small space with a big bang! In 1948, this idea was given substance in a famous paper by Alpher, Bethe, and Gamov (Gamov chose the other two authors as a pun on the first three Greek letters, alpha, beta, gamma). They pictured an early universe at such a high temperature that only radiation could exist. Its subsequent history, using present day laws of physics, has been much theorized and discussed.

When the universe had cooled sufficiently, protons and neutrons began to form and these subsequently combined to form helium nuclei. A few hours after the bang, helium production had stopped, and soon it was cool enough to form electrons which combined with nuclei to form atoms. Eventually atoms coalesced into clouds which in turn collapsed into stars and galaxies. When angular momentum was present in the cloud, the result was a spiral galaxy. If none was present, elliptic galaxies were formed.

The intense pressure and renewed high temperatures of the newly formed stars led to nuclear fusion and the formation of atoms more massive than helium. Calculations of the estimated temperatures and pressures in the early stars, together with known formation rates of nuclear species, yield abundances of the 92 elements in rough agreement with estimates of their abundance in the Sun.

In 1965 two physicists from Bell Laboratories, Penzias and Wilson, checked the prediction of Gamov that the Big-Bang fireball should have cooled to about -270° C by this time. They measured the temperature of radiation from outer space, using directive antennas at microwave radio frequencies, and roughly obtained that temperature. This convinced many that the Big-Bang theory was true.

There are many unanswered questions concerning the Big-Bang as in Figure 11-2. The theory has no way of explaining what happened *before* the Big-Bang. The "bang" describes the instant when volume is zero and temperature is infinite. This situation, termed a mathematical *singularity*, is well known in physical analysis to lead to an indeterminate situation where little is certain. The necessary assumption that the laws of physics were unchanged during a process in which the *entire matter of the universe* began at or near infinite density, temperature and zero volume is difficult to believe. In order to believe it, one must set aside any convictions that physical laws depend on gross properties of the universe itself.

THE BIG BANG THEORY – A visual interpretation by illustrator

And lo, amidst total chaos & violence the universe appears, all 10^{80} protons of it, squeezed into zero volume, with all fundamental laws intact.

Then, stars form, and spread apart like a perfect raisin pudding: fresh from the oven...

OUR LAWS OF PHYSICS ARE IMMUTABLE FOR ALL TIME. ALL ENERGY IS CONCENTRATED IN ZERO SPACE.

SO WITH *THAT* LOGIC, WE SHOULD BE ABLE TO PREDICT 100 BILLION YEARS IN THE FUTURE.

WHAT HAPPENED BEFORE T=0?

BIG BANG THEORY AS IT IS PRESERVED TODAY:

SUDDEN EXPANSION

MULTIPLE WORLDS

R.I.P.

© 90 SNOWOLFF

FIGURE 11-2. Theory of the Big-Bang.
According to one interpretation of Hubble's measurements, all the stars and galaxies are steadily moving apart, one ten billionth part each year. This suggests that 10 billion years ago, all of the matter of the universe occupied a space of zero volume. This implies, using the laws of thermodynamics of today, that the temperature at time zero was infinitely hot, and the density was infinitely large. This remarkable conclusion and the theories which interpret it are termed the "Big-Bang."

There is no theory at all about the universe before the Big-Bang. You are not supposed to ask that question. However, many calculations of the behavior of matter after the Big-Bang have been published, and they predict, of course, that our world today is just like it is.

There are philosophical objections, too. The predictions of the Big-Bang which refer to conditions of the present time must be suspect, since the answer is known before the calculation is made. One could also ask, "How much confidence do we have in a prediction using Big-Bang theory, concerning the state of the universe 10^{10} years in the future? The Big-Bang theory is in conflict with Tetrode's Principle (discussed in Chapter 14), since if Tetrode is correct, there can be no radiation without pairs of matter that exchange it. Thus the assumption that the Big-Bang consisted only of radiation is contradictory. Of course Tetrode's idea is unproven, too. Nevertheless, there is no room in the Big-Bang for the idea that photons are an energy exchange with matter; a process we don't understand anyway.

One of the most attractive features of the Big-Bang is that it is a fruitful source of material for publishing papers. It has also stimulated thinking about alternatives such as the not too serious suggestion of Figure 13-3 in Chapter 13.

"The evolution of great ideas has always been opposed by mediocre minds."

— Albert Einstein

CHAPTER 12

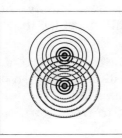

The Space Resonance Theory

An Overview of the Space Resonance Idea
The First Assumption of the Theory
A Closer Look at the Strange IN-Waves
The Second Assumption of the Theory
The Third Assumption of the Theory
Charge and Mass Paradoxes Resolved
Nuclear Resonances: the Same Space but a Different Resonance
The Weak Nuclear Force: The Same Resonance but a Different Space
Tetrode's Speculation

CHAPTER 12
The Space Resonance Theory

An Overview of the Space Resonance Idea

At the end of Chapter 10, the six fundamental laws and three particles which determine the rules of science and make up the matter of the Universe were identified. We want to understand the origins of the six laws and we need a model of the particles, free of enigmas and paradoxes.

In this chapter I propose a model of the structure of the three basic particles. Three assumptions about space properties are used. The behavior of the model reveals an origin of four of the laws: quantum mechanics, relativity, electric charge, and conservation of energy. The remaining force laws, gravity and inertia, result from disturbances (perturbations) of the electric force due to the Hubble expansion and acceleration of matter.

The idea for this model began with a speculation that waves in space could explain the deBroglie wavelength. When I added a mathematical description of the model's waves, it agreed with other laws, and analysis showed it could explain many paradoxes. This surprised me so I continued to work out more consequences of the model, but it is far from complete.

Will this concept turn out to be correct? Experimental tests are proposed, but you and other readers will be the judges. If the concept is correct, we will have found an important basis of science, and I hope that you will continue to explore it beyond this frontier.

You have already guessed that the model uses waves in space. But the model uses no charge or mass substance; just empty space which we ordinarily regard as "nothing". According to this proposed model, properties of space are the basis of all physical laws, matter, and the structure of the universe.

I have named this particle structure a *space resonance (SR)*, because a common term for a standing wave is a *resonance*, and because these waves propagate in the medium of *space*. According to this concept, all the matter of the universe is comprised of space resonances whose waves travel in space usually without disturbing each other.

If space resonances exist, we logically expect that of their properties derive from properties of space because it is a standard result of wave theory (Chapter 8), that all wave properties arise out of their media. But I did not anticipate this since I had a *particle* model in mind at first, not just waves. Only in hindsight after the theory was mostly complete did I notice that the fundamental properties I had found were all properties of space. This discovery was exciting because it indicated that the theory was consistent, a necessary condition.

Below I will analyze the space resonance properties and it will be seen that the laws of quantum mechanics and special relativity are a result of the motion of a space resonance relative to any other resonance. The force laws between particles are found by postulating two assumptions: one, a process of energy transfer, and the other, a principle that motivates energy transfer between space resonances. Thus three assumptions are needed to obtain the space resonance and its properties.

Recent Attempts to find the Structure of Particles. The desire to understand the fundamental laws identified in the preceding chapters has been an endeavor on the forefront of scientific research. There is a special attraction for this work because the laws are universal, ruling the behavior of all matter. No one has understood these foundations of science, but many scientists have had a gut feeling that somehow the unknown fundamental laws are tied together, so that if we could only grasp a few underlying concepts, we would then gain insight into the whole group.

Many ideas have been tried to find a hidden universal concept. One favorite hope is to unify some or all of the four force laws; gravity, electric, nuclear and weak forces. Another idea is to locate the smallest indivisible building block to form all the particles. Engineering-oriented scientists often try to make models of particles using electromagnetic waves, confined to small spaces, so that they possess the observed particle properties as well as the wave properties. Mathematical particle models attempt to relate the terms of their equations to building blocks. These attempts have all run into difficult barriers.

A few attempts have been made to intimately relate micro-particle structure to cosmology. Many are similar to the philosophical suggestions of Ernst Mach, Paul Dirac, and Tetrode, mentioned in Chapter 11, but no experimental means has been found to verify the cosmological relations. Some of the ideas of Einstein's general relativity have been turned into very imaginative entities or "particles of cosmological space" including black holes, white holes, and worm holes.

The Requirements of a Wave Particle Model. Before examining my new particle model, look at some of the requirements it must satisfy. Two generic properties are needed to identify it as a particle. These are,

1) It must have *location*. That is, you must have a way to know it is "here," not "there."

2) It must have one or more ways to *exchange energy* with other particles. For example, an electron has two ways to exchange energy: by an electric force and by a gravity force.

A third necessity for any *real* particle model, which most everyone knows, is that its properties must contain the laws obeyed by the electron, neutron, and proton — the actual citizens of the particle world. This necessity adds two more requirements, namely:

3) It must obey the laws of quantum mechanics (the de Broglie wavelength, and the wave-probability concept).

4) It must obey the laws of relativity (increase of mass due to relative velocity). And finally a fifth requirement:

5) A particle model must obviously match the special properties of the particle itself.

A Quick Look at the Space Resonance. The theory proposes that spherical wave resonances occur in the fabric of space. The waves are scalar waves, not electromagnetic waves, and their center is the location of a particle. These resonances are proposed to be the fundamental particles: electrons, protons, and neutrons.

From this model we can find the laws of quantum theory, the relativistic change of mass, the properties of an electron, its charge, the exchanges of energy with other charged particles, and all the electric and magnetic properties that are derived from moving charges. The simple "machinery" of the model leads to a clear understanding of *how* the laws work, dispelling the paradoxes of QM and relativity.

The model machinery allows you to envision particle behavior, to deduce the laws of particle interaction, and to see how the properties of the Universe depend upon the particle properties and vice-versa. It differs from other explanations of frontier physics which tend to be very mathematical and don't have machinery which tells you "how."

The model consists of two spherical waves with the same center. One moves convergently towards the center and is termed an "IN" wave. At the center, it reverses direction to become an "OUT" wave which moves divergently away from the center as shown in Figure 12-1. Since the two waves are of the same frequency, they form a standing wave, which if you could see it would resemble

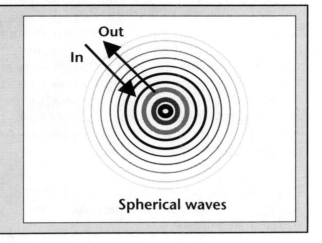

A Space Resonance Has Spherical Symmetry

Spherical waves

FIGURE 12-1. The moving waves of space resonance.
The resonance is composed of a spherical IN wave which converges to the center and an OUT wave which diverges from the center. These two waves combine to form a standing wave whose spherical peaks and nodes are like layers of an onion. The amplitude is a scalar number, differing from the electromagnetic waves which have a vector amplitude. At the center, the standing wave amplitude is finite rather than infinite, in better agreement with electron measurements than the classical point electron model.

the spherical layers of an onion. The amplitudes of the standing waves decrease with distance from the center in inverse proportion to the distance; varying as $1/r$. Ordinarily, the waves of each resonance pass through waves of other resonances without interference so that all space is filled with waves moving in every direction into and out from all of the particles in the universe. The simplest resonance has the properties of an electron.

The Spherical Waves are Solutions of a Scalar Wave Equation. The usual way that Nature communicates the rules of waves to scientists is in the form of a wave equation which frequently expresses an underlying principle, often concerning balanced forces or energy relationships. Since waves in Nature are always moving, the equation has to be a relation between derivatives with respect to time and space that express movement. Solutions of it will be an amplitude in space moving at a given time rate.

Historically, wave equations have usually been guessed at first and the principle discovered later. I have done the same by assuming a scalar equation to determine the amplitudes of the waves as a function of time and space. Why space possesses this wave property is quite unknown, but it seems to work.

Certain clues are available to help the guessing process. First we need a wave distribution with spherical symmetry that will match the symmetry displayed by charge and mass forces of particles. Vector wave equations were ruled out because they have been extensively explored, especially the electromagnetic wave equation, and it is already known there are *no* e-m wave solutions which are spherically symmetrical. Also, we desire a connection with scalar quantum waves and this suggests that a scalar wave equation may be suitable. For geometric reasons (for example, properties in the +x direction should not be different than those in the -x direction), a wave equation should have second derivatives in the space coordinates.

The First Assumption of the Theory

Using these clues, and choosing the simplest relation between time and space derivatives, the first assumption of the theory is a scalar wave equation as follows:

WAVE EQUATION ASSUMPTION (I)
Space can propagate scalar waves according to

$$\nabla^2 \Phi - (1/c^2) \, \partial^2 \Phi / \partial t^2 = 0$$

where Φ is a continuous scalar amplitude with values everywhere in space, and c is the propagation speed.

This equation is similar to many other oscillatory equations found in nature. It is a propagation equation for the particle waves and provides the law for their formation and structure. The wave amplitude as a scalar quantity has only a single numerical value at each point in space. Thus, it is not a vector quantity like light waves which have both amplitude and direction.

It is evident that this equation of *space resonance waves* is an implicit assumption that space has the properties needed to propagate them. Those properties are not yet understood.

This equation is solved using spherical coordinates in the Appendix if you wish to read it. One solution is an incoming spherical IN wave, and another is an outgoing spherical OUT wave. These two may be added to each other to form another solution which is the standing wave mentioned above. The result for the amplitudes Φ of the standing waves, taken from the appendix, is:

$$\Phi = \Phi_0 \, e^{i\omega t} \sin(kr)/r \qquad (12\text{-}2)$$

The first factor Φ_0 is the maximum amplitude. The exponential factor is an oscillator of frequency ω. It is modulated by the sine factor, which is a standing wave of wavelength **1/k**. The amplitude becomes smaller as the radius **r** increases because of the **1/r** factor.

The Properties of a Simple Space Resonance Look like an Electron. Analysis of the algebraic structure of the wave function Φ shows properties which match those of particles such as electrons, as follows:

1. Spherical Waves have Spherical Symmetry. The spherical symmetry of the space resonance waves agrees with the observation that particle forces have spherical symmetry. The fact that spherical solutions exist, contrasts with the electromagnetic wave equation which has no spherical wave solutions.

The amplitude of the waves decreases with distance **r** exactly like the forces of charge and mass. Charge and mass forces are introduced into the model below with two more assumptions about properties of space. Charge and mass are properties of space and the waves, rather than substances carried by the particle.

2. The Wave Amplitude is finite at the Center. The standing wave amplitude at the center is Φ_0 as you can see if you put r = 0 into the above equation for the amplitude. This feature resolves a paradox of the usual electron model, because the usual potential energy (PE = q/r) of a point electron's charge becomes infinite at r = 0, which is obviously impossible. This unsatisfactory infinite field is a well-known difficulty of the usual electron model. Instead, a finite amplitude at r = 0 is necessary to satisfactorily explain electric charge behavior. Surprisingly, the space resonance provides it.

3. The Space Resonance has an anti-resonance. The wave solution shows that space resonances come in two varieties. The second variety is an anti-resonance, just like an electron which has an anti-particle, the positron. Two wave solutions occur because of the three-dimensional geometry of space. In one solution, the "IN" wave amplitude is positive at the center and the "OUT" wave is negative. In the other solution, the OUT amplitude wave is positive and the IN is negative. Both the resonance and anti-resonance standing wave solutions have the same algebraic structure except for opposite signs of the amplitude; that is, they have a phase difference of π radians and are related by,

$$\Phi^{(+)}(r) = -\Phi^{(-)}(r)$$

As a result, if a resonance approaches and is superimposed upon an anti-resonance, their opposite amplitudes annihilate each other, just as the electron and positron can annihilate.

4. The wave node spacing is the Compton wavelength. If the frequency $\omega = 2\pi f$, of the resonance is chosen to be the electron mass-frequency, $f = mc^2/h$, then the distance between wave nodes is h/mc, the *Compton wavelength*, already a familiar quantity in measurements of electron behavior and theory.

Previously, this quantity had no physical interpretation; it was just a mathematical length that could not be identified with any property of the electron. Now this length turns out to be a dimension of the resonance.

We next examine the effects of motion, and we shall see that the resonance and the anti-resonance also have properties like a particle and an anti-particle in other ways, including charge, time reversal, relativistic mass-change, and deBroglie wavelength.

Relative Motion of two Resonances. In 1950 when I was a student at the University of Pennsylvania, I wondered what was the origin of the deBroglie wavelength given by $\lambda = h/p$. It is such a strange relation being inversely proportional to the momentum p. There is absolutely nothing physical about it, because intuitively you expect small wavelengths from small momentum, but no, it is just the opposite. I searched all my textbooks for an explanation of where it came from, but there were none. The subject was never discussed in class, but I thought I had missed an obvious point, so I said nothing despite its clear importance (all of quantum theory grows out of it). This problem had nagged me until later I saw an answer in the behavior of two moving space resonances, which I will now describe.

We know that relative motion is the logical beginning of both the laws of relativity as well as quantum mechanics. So, it is natural to ask what happens when two space resonances are moving with respect to each other. Since they are waves, each must see a Doppler change of the other's wavelength. To find how much, you must write the equations for them and calculate the Doppler effect. This is done in the appendix and it is found that two moving resonances surprisingly display both the relativistic and QM properties of a particle.

Logically it does not matter which resonance is observing the other. Both resonances must see the same appearance in the other because the configuration is symmetrical. Neither one can be said to be fixed and they each have the same relative velocity. This symmetry is observed in relativity and QM measurements, and appears in the equations, but there has been no physical explanation of how it arises. Now it can be seen to be a Doppler frequency shift of the resonances.

To begin, let us find the apparent changes in the waves of a moving resonance when they arrive at another resonance. For purposes of clarity, the frequency of the resonances will be chosen the same as the mass-frequency $f = mc^2/h$, of a particle such as a proton or electron. By doing this the relationship to a real particle can be more clearly seen.

Since the resonances have relative motion with velocity $\beta = v/c$, they will each see each other's IN and OUT frequencies changed due to the Doppler effect. Both resonances receive the same information from each other because the relative velocity is the same. The calculations are made in the Appendix, and the resulting amplitude for the Doppler shifted wave of either resonance, which arrives at the other resonance is

$$\Phi = 2 A e^{ik\gamma(ct + \beta r)} \sin[k\gamma(\beta ct + r)] \qquad (12\text{-}3)$$

where **A** is the wave amplitude which includes the **1/r** factor, **k** = mc/h, and γ = $(1 - \beta^2)^{-1/2}$.

This equation has the form of an exponential carrier wave modulated by a sinusoid. The surprising parameters of the carrier wave are:

>**wavelength = h/γmv = deBroglie wavelength**
>
>**frequency = kγc/2π = γmc²/h = mass-energy frequency**
>
>**velocity = c/β = phase velocity.**

The modulating sine function has:

>**wavelength = h/γmc = Compton wavelength**
>
>**frequency = γmc²β/h = b x (mass frequency) = "momentum frequency"**
>
>**velocity = βc = v = relative velocity of the two resonances.**

These wave properties are the same as what you get from a particle using the laws of special relativity, and quantum mechanics.

QM and Relativity Appear in the Waves. All the parameters that can be measured for a moving particle, are correctly contained in the above equation. Each resonance receives from the other, two quantum-mechanical parameters: a deBroglie wavelength and the Compton wavelength; and the relativistic *four-momentum* vector: rest mass and three components of linear momentum. I had not expected to find this. The discovery of this result in 1983 was an exciting event in my life, because it provided a possible origin for the elusive deBroglie wavelength! This possible origin was strengthened by the unexpected presence of the relativistic parameters.

There is another unusual thing about these properties of the space resonance. If you play around with the equations of the waves, you discover that, in order to get those results (the deBroglie wavelength and relativistic mass-change) the IN and the OUT waves are both absolutely necessary. There is no method to re-arrange either one alone, or even just make up some equations to do it. The IN and OUT waves are the heart of it all. This implies that these two most fundamental laws of Nature are a result of two interacting spherical waves; one going in and one going out. If this idea is correct, there is no longer anything perplexing about QM and relativity except for the IN wave, which although very simple, is also puzzling.

A Closer Look at the Strange IN-Waves

Everything in the above wave equation and its mathematical solution are understandable and thus undisturbing, except the IN waves, which are very puzzling because they seem to go in the wrong direction. You are immediately entitled to ask, "Can they be real?" They arise at infinity with zero amplitude, grow larger, and then converge at the center point. They have the same wave properties as the OUT waves except that the latter begin at the center point and go out. It appears almost impossible that the IN waves know where they are going before they get there. Are they moving backward in time?

Of course, you can reason with yourself and possibly convince yourself, that neither IN nor OUT wave is more strange than the other. Their symmetry allows an argument about one to be applied to the other. Our human realm familiarity with outward going waves causes us to be prejudiced against IN waves.

The IN/OUT wave combination is the heart of a resonance. Yet, we will see that the IN waves, in combination with the OUT waves, are essential to the entire concept of the space-resonances. Without them the entire idea collapses. If you carefully follow the mathematics of the wave-equation solutions in the Appendix you conclude that it is the mixing of these two waves which creates the new features which underlie and explain quantum mechanics and relativity. Both waves are needed.

In quantum mechanics, both waves are needed to form the DeBroglie wavelength, the most basic element. Recall that according to QM the deBroglie wavelength does not depend on whether particles are moving together or apart, it is symmetrical with respect to direction just like the IN/OUT waves. Later, other unusual features of quantum theory will be seen to focus on the IN/OUT wave combination.

In relativity, recall that mass increase takes place symmetrically when a particle is moving either towards or away from the observer. No classical explanation has ever been found for this symmetry. But now, it is explained by the spaces resonances, where it arises from the symmetry of the IN/OUT waves with respect to time and space between the resonances.

Where is the IN/OUT Wave's Road Map? There are two ways to explain how the IN/OUT waves maintain their identical wavelength and common center, despite their apparently different sources. One can speculate that they continuously exchange wave information as they move past each other. The exchange is caused by minute irregularities in space which act as tiny non-linear elements. Each wave copies the other in passing. This graininess must be very minute, since the number of particles in the Universe is very large — about 10^{80}. Therefore this argument is uncertain.

It can also be argued that there is no need for communication between IN and OUT waves, because space may only propagate *one* type of wave, which has one wavelength, one frequency, and one propagation velocity, which depend only on properties of space which are universal.

These ideas seem reasonable, but there is still the troubling matter of how the IN waves know where they are going, especially if the center is in motion or under acceleration. This puzzle also exists in a subtle way for an OUT wave under acceleration. A lot of thinking and exploration still needs to be done!

It is some comfort to realize that similar problems already exist for our explanations of electromagnetic waves, only much worse. Illogically, most everyone accepts e-m waves without asking questions, because we feel that we observe them, especially if they are light waves. But do we observe them?

The last comfort is that if you can accept the strangeness of the IN/OUT waves, you will be rewarded by no more strain on your credulity — because then everything else falls into place reasonably. The IN/OUT waves explain other puzzles of physics, leaving the waves as the only remaining puzzle.

New ideas after an enthusiastic inception often prove to be false so one has to check carefully to sift out the good ones. To verify quantum mechanics, to find an origin of the deBroglie wavelength is not enough; one also has to ask whether the rest of quantum mechanics can be satisfied? To do this, the space-resonance theory has to include the QM probability functions $\Psi^*\Psi$. This will be discussed in the next section after two additional assumptions are introduced.

Feynman said, "A positron is an Electron going backward in Time!" The gurus of science like to startle their students by saying things which are clearly true but seem strange or impossible. Noble Laureate Richard Feynman in his lectures on QED stated that you can draw a positron in a "Feynman diagram" as an electron going backwards. It always works out right, but no student has ever found a physical reason why. How can a particle go backward in time?

The space resonance shows you how. If you go to the equation for an electron space resonance in the appendix and replace the variable t with a -t, the result is just the equation for a positron! Physically the change of time direction means that the IN wave is changed to an OUT wave and vice versa. Their roles are reversed, including the + or - amplitude at the center. To understand this you don't have to imagine clocks running backwards. Just exchanging the IN and OUT waves produces Feynman's statement.

The Second Assumption of the Theory

When quantum theory was first begun, it was soon realized that if a particle was represented by a wave, the assumed location of the particle was not definite as the Greeks had imagined. You could not point to a certain place in a coordinate system and say, "The particle is there." Instead, you had to say, "The particle is probably near this place where the waves are most intense." Similarly, you could not exactly specify a momentum, or energy property dependent on the waves. To calculate these probabilities in quantum theory, we use the mathematical expression Ψ, which describes the amplitude and intensity of the waves. Then $\Psi^*\Psi$ is used in various ways to find the probable values of position, momentum, and energy.

In order to prove that space resonances are related to quantum theory, we need to show that space resonances also contain a QM probability function like Ψ, such that

$$\text{probability} = \Psi^*\Psi \, dxdydz \qquad (12\text{-}4)$$

which represents the probability of "finding" a particle in the volume of **dxdydz**. To "find" a particle means to carry out an energy transfer since only a transfer can provide the data of measurement. So far, the wave model has no means of energy transfer. There must be an energy transfer mechanism for the space resonance model; otherwise it fails.

To find the mechanism, I looked for a *non-linear* region of space, because propagation through it is the only known means to transfer energy (shift frequency) between two sinusoidal waves. Such a region could exist at the center of the resonance where the wave density is greatest. So I assumed that wave propagation depends on the density of space, and that space is not homogeneous at the center of the space resonance.

> **SPACE DENSITY ASSUMPTION (II)**
> The propagation of particle waves is a function of the density of space. And, the density of space is dependent upon a super-position of the densities of waves from all particles inside the Hubble horizon of radius $R = c/H$, including the density of a particle's own waves.

To express this idea quantitatively, write

$$m_m c^2 = h f_m = C_m \Sigma \Phi_n^2 (1/r_n^2) \qquad (12\text{-}5)$$

which means that the frequency (mass), of a particle **m** depends on the sum of intensities Φ_n^2, of all waves Φ_n, from the **n** particles in the universe, whose intensities decrease inversely with range squared, and on the type of resonance c_m, of the particle **m**.

In most of space, the total intensity from all the particle waves is almost constant, simply because of the large numbers of particles in the Universe, so that the propagation speed **c** is almost always constant as observed. But at the center region of a space resonance, according to (12-5), the space density must be significantly increased because of the large intensity of the resonance's own waves. This results in a non-linear condition at the central region and is assumed to provide the mechanism needed to allow energy transfer. The non-linear condition has an analogy in electric circuits where it is well known that if a circuit element with two signal currents flowing in it is non-linear, the result will be cross-modulation, or a partial mixing of the two signals, a frequency transfer.

Is This A Reasonable Assumption? Can this idea be tested? The final test of this assumption is whether or not it works. A rough test calculation can be made to see if this assumption is reasonable, because the radius **R**, of the visible universe and the total number **N**, of particles (mostly hydrogen) is roughly known. If a resonance's own waves are to have any significant effect on space density, then, according to (12-5) the intensity of the particle's own waves at some radius r_o, must be roughly equal to the total intensity of all waves from all the other **N** particles in the Universe. If this is not true, the assumption must be wrong. This test condition can be expressed,

$$I(\text{own wave}) = \Phi_o^2/r_o^2 \approx \Sigma \Phi_n^2/r_n^2 = N/V \int_0^R \{\Phi_o/r\}^2 4\pi r^2 dr \qquad (12\text{-}6)$$

where **V** is the volume of the universe and **R** its radius. Condition (12-6) reduces to

$$r_o^2 \approx R^2/3N \qquad (12\text{-}7)$$

Inserting values from cosmology (Berry, 1976): $R = 10^{26}$ meters, $N = 10^{80}$ particles, one obtains a value of the critical radius $r_o \approx 0.58 \times 10^{-14}$ meter. One would expect this radius to correspond, at least roughly, with the classical radius $r = e^2/mc^2$ of an electron. That radius is, $r = 0.28 \times 10^{-14}$ meters. So, within the errors of the cosmological data, the test condition is satisfied, and the assumption passes the first test.

A Connection Between the Biggest and the Smallest. But, this test gives us something more! The condition (12-7) from the *Space Density Assumption* also provides a numerical relation between cosmology and microphysics. The cosmological dimensions, **R** and **N** of the universe, are equated to micro dimensions, r_o of the electron. This relation is an answer to a research goal of several decades standing. The relation is now crude because **R** and **N** are not well known, but when the measurement errors of **R** and **N** are decreased, this equality may be a confirmation of space resonances.

Three Conditions are needed for Energy Transfer. The energy transfer (frequency change) mechanism for the space resonances is furnished by the *Space Density Assumption* which hypothesizes a non-linear condition at their wave-centers. But in order for a transfer to occur, three conditions must be present:

1) A non-linear region must exist at the wave-centers to allow an exchange of energy (frequency change) between waves.

2) The waves of each interacting resonance must overlap the wave-center of the other.

3) A driving condition must exist. The first two conditions provide the means for an exchange of energy (change of frequency), but they do not drive it. A drive is needed, for example, to determine in which direction an energy exchange should go.

Without all three conditions, particle waves travel independently in space without interaction. The third driving condition is provided by the *Minimum Amplitude Principle,* discussed below.

The Third Assumption of the Theory

The condition needed to bring about an energy transfer process is a motivation for the exchange; a final equilibrium state towards which space resonances tend to move. If they are not in the final state, their non-equilibrium will cause them to move towards it. This kind of "compulsion by Nature" is frequently invoked by *Principles*, such as the *Least Action Principle* and the *Increasing Entropy Principle*. That compulsion is embodied in the following assumption:

> **MINIMUM AMPLITUDE PRINCIPLE ASSUMPTION (III)**
> The total amplitude of all particle waves in space always seeks a minimum at every point. As a result, the centers of space resonances move, accompanied by frequency changes (energy exchanges), so as to approach the minimum value.

This Principle (MAP) describes a property of space which tends to minimize the total amplitude of particle waves. One can compare this to the tendency of water in the ocean to seek a minimum level. Accordingly, whenever the waves of two resonances overlap and are able to change frequency (exchange energy) to reach a configuration of lesser total amplitude, they will do so. When wave amplitudes overlap (and add), space density is increased, and frequency is increased according to equation (12-5). If wave amplitudes cancel, space density is decreased and frequency is decreased.

Minimization can be expressed as an integral:

$$\iiint [\Phi_1 + \Phi_2 + \Phi_3 + \ldots]^2 \, dxdydz = \text{a minimum}$$

Energy (frequency) changes take place in order to adjust the total wave amplitudes to the minimum sought by the Minimum Amplitude Principle and simultaneously wave-centers move to achieve the configuration of minimum wave amplitude.

The Energy Exchange Process Reveals two Fundamental Laws. The main reason for the assumption of an energy (frequency) exchange process was to provide a connection with quantum mechanical probability. But as occurred before, two unexpected extra benefits appear.

1) First, this exchange process also turns out to explain the *law of conservation of energy*, as follows:

Conservation of Energy. Two space-resonance oscillators exchange energy much like classical coupled oscillators, such as electric circuits or joined pendulums. The coupling provided by the non-linear centers of the resonances allows them to shift their frequency patterned by the modulation of each other's waves. Since significant coupling can only occur between two oscillators which possess the same resonant elements, the frequency (energy) changes are equal and opposite. This we observe as the law of conservation of energy.

When opposite changes of frequency (energy) take place between two resonances, energy seems to be transported from the center of one resonance to another. We observe a loss of energy where frequency decreases and added energy where it increases. The exchange appears to travel with the speed of the IN waves of the receiving resonance which is c, the velocity of light. When large numbers of changes occur together so we can sample part of it, we see a beam of light.

When single exchanges occur we see "photons." The transitory modulated waves traveling between two resonances create the illusion of the "photon." An exchange may require 10^8 to 10^{15} cycles to complete, depending on the degree of coupling and species of resonance. For example, if one oscillator were an electron, its frequency mc^2/h is about 10^{23} hertz, and if the transition time were 10^{-8} seconds, the frequency change requires about 10^{15} cycles to complete. Such a large number of cycles implies, in engineering slang, a large Q value, which indicates great precision of the equal and opposite changes in oscillator frequency, and the conservation of energy.

(2) The second extra benefit is a unified origin of all the forces of Nature, as follows:

The Unified Origins of Forces. Force is defined as the energy change (frequency change) per unit distance moved $F = dE/dx$. Experimental scientists have found the rules to calculate Nature's forces. But neither the cause of the forces nor the origin of the constants are known. The experimentally found numerical constants are what we call "charge" and "mass" and nuclear binding energy.

According to the space resonance idea, forces arise from the adjustment of wave amplitudes caused by the *Minimum Amplitude Principle*. The frequencies of interacting oscillators (particles) shift oppositely accompanied by relative movement of the particles. When this occurs, the force we observe is the ratio between the frequency shift Δf and the accompanying relative movement Δx. Thus force = $h\Delta f/\Delta x$. Since all forces arise in this same way, it is a *unified origin of forces*. In principle, the force constants could be determined from the properties of space when we measure and learn more about it.

Quantum Mechanical Probability. An energy transfer mechanism is now established and we can return to the question of whether or not the wave model incorporates the probability concepts of QM.

To see the connection with QM, examine a measurement process where one space resonance (SR) is used as a detector to find the location of a second SR whose waves are described by a function Ψ. Assume the two are not in a minimum-amplitude state, so that energy can be transferred. The rate of transfer of energy to the detector resonance will be proportional to a constant describing the detector wave-center and to the wave intensity $\Psi^*\Psi$ of the second (measured) resonance. If the detector SR is moved around in the wave field of the measured SR, the rate of energy transfer will map out the $\Psi^*\Psi$ function. Hence the SR energy transfer mechanism corresponds to QM probability functions.

Correspondence with QM can be seen by another example. Examine the energy exchange between two states Ψ_1 and Ψ_2 of two different resonances. The function $\Psi_1^*\Psi_2$ is a real number proportional to the overlap of the two wave sets, which is proportional to the rate of transfer. But this is the same as the integral over all space of the two sets of waves

$$\int^\infty \Psi_1^* \Psi_2 \, dxdydz$$

which is the QM matrix element between states 1 and 2; which is just the QM probability of a transition between them. So the SR and QM viewpoints are proportional.

In summary, it is seen that the QM probability functions $\Psi_i \Psi_j$ appear to have their basis in the SR energy-transfer mechanism, for which the *Minimum Amplitude Principle* provides a motivating force.

Charge and Mass Paradoxes Resolved

The word *charge* must describe the same observed forces for both classical particles and space resonances. However, the way of thinking about a SR and a classical charge are quite different. These differences resolve the paradoxes associated with a point of charge or mass.

Is Charge a Property of a Particle or Space? Beginning with the early experiments of Alexander Volta and Benjamin Franklin, charge was always imagined as something which flowed along metal surfaces and could be kept in special bottles. The early capacitor was called a Leyden jar and in dry weather the charge could be kept for days, "in the jar" like a genie of the Arabian Nights. Indeed, the flow of electricity does mimic in many ways the behavior of a liquid in a pipe and this analogy is often used in teaching. Accordingly it is not surprising that a concept grew up in which charge was regarded as a mobile substance possessed by a particle.

Classically, charge is the numerical value of the constant e in the electric force equations. It is a mysterious substance that is conserved, associated with particles, but never directly observed or isolated. The concept of an electron has been almost completely associated with the charge and mass substances so that its structure as a particle arose from the need for a vehicle to carry charge and mass from place to place. Disturbing this tidy arrangement, its mathematical force $F = e^2/r^2$ leads to a paradox predicting infinite energy when r = 0, which is impossible. The same infinity paradox occurs for gravity or for inertia, in a different mathematical context.

It takes Two to Tango. The scientific evidence does not support this popular idea of the electron. There are no laws of physics in which e appears alone. Instead the laws always involve *two* charged particles and any phenomena relating them is proportional to both charges, so that charge always appears as the value e^2. Thus the laws do not substantiate the charge "substance" existing on the electrons. Unfortunately most textbooks still retain the substance charge e and ignore the fact of e^2.

There is other evidence. There are no particles of any sort which possess charge *properties* alone. Charge is always accompanied by other properties including *mass*. It may be a coincidence but the geometric dependence of mass and charge forces are both the same, being proportional to $1/r^2$, distance between the particles. The energy exchanges of both forces are communicated with the speed of light through the space between the two particles.

In contradiction to popular opinion, we must conclude that charge is a joint property of both particles (space resonances) and of the space around them. It does not make sense to speak of charge as a property of *one* particle; instead, it is an energy exchange property between *two or more* particles, which can be expressed as a force between two particles, $F = \Delta E/\Delta x = h\Delta f/\Delta x$. To sum it up, "It takes two to tango!"

If I can't see it, it is probably a point! This is a typical human reaction to something tiny. In hindsight it is easy to understand how this mistaken notion of an electron arose. The wave properties of an electron are extremely small compared to early instruments. The center wave diameter, of about $h/mc = 10^{-12}$ meter, is so small that during the first two centuries of electrical research it appeared to be a point. Even today, measurements that small are rare indeed.

One Space Resonance Does it All (Charge and Mass Included). In a space resonance, the notion of a "particle," to carry mass and charge substance, is not necessary. Instead *charge, mass, and all other properties of the particle are implicit in the behavior of the space resonance and its wave fields.* Discarding the particle as a vehicle produces important conceptual differences. Classically, the QM waves, mass and charge are separate entities. But in the SR they are one entity where all properties derive from the space waves. This difference between the old point particle and a SR resolves all the paradoxes associated with the point particle, including:

1) the wave-particle duality of QM (Which hole did the particle go through?)

2) the infinite energy at r = 0, and

3) second normalization of quantum electrodynamics.

The SR concept resolves the paradox by removing the need for vehicles to carry energy, mass and charge. These old "packhorse" roles are not necessary because the energy exchange, mass, and charge are part of the behavior of the SR waves themselves.

The Photon and the Wave-particle Duality Paradox. The central paradox of quantum theory (Chapter 9) is the apparent duality of particles which sometimes behave as particles and sometimes as waves. It is now easy to understand how the wave-particle duality paradox arose. It was spawned by the notion that point-like mass and charge were the substance of the electron. These points must somehow travel inside the envelope of quantum waves, then finally and magically appear "somewhere" whenever the electron was detected.

This paradoxical duality does not exist for the space resonance since there are no point-like substances to be carried along.

Part of this paradox is related to the photon, which is assumed to be a carrier of the energy exchanged between two charged particles and thereby produces another image in our minds of a "particle." Having created the particle images, our mind demands that they be accounted for and our inability to do this, because there are no particles, brings on the paradox. To resolve the paradoxes we needed to remove the particle idea from both the electron and the photon.

The Machinery of Energy Transfer is an Interaction of Two Oscillators. Review the mechanism of the SR electron charge property. It is due to: 1) The *Space Density Assumption* which results in non-linear space when the wave amplitude is large, thus, allowing the electrons to modulate each other's waves. 2) The *Minimum Amplitude Principle (MAP)* which causes the two electrons to readjust their frequencies seeking a minimum total amplitude as the distance between them is changed.

Using these two principles, the energy transfer between two relatively moving electrons is easily understood as an interaction of two oscillators. Their wave-sets seek a minimum total amplitude by moving the wave-centers together or apart with frequency changes (exchanges of energy) as they move. Each movement creates a traveling modulation of the waves to and from each SR. This temporary modulation is the phenomenon that corresponds to the traveling photon. The process is very close to the behavior of two coupled oscillators in an electronic circuit or even two pendulums joined with a coupling spring. But in this situation the coupled oscillators exist as properties of space.

Is the Photon Needed? The photon concept seems unnecessary, even confusing, if we can believe that the energy exchange between particles is a frequency change of the two resonances brought about by a modulation of the particle waves. The photon is not needed since there is nothing to be "delivered" between particles.

The photon paradoxes are partly psychological because our minds seek object-like images from the data they receive. Light sources appear like beams and we imagine that we see a ray of light traveling from here to there, but in reality, there is no possibility of observing energy in transit. It is an illusion, usually the result of dust in the light path! Photocells yield pulses when their single atoms are stimulated, so our visual apparatus imagines an arrival at the receiving places. These appearances lead us to believe that the energy is carried by traveling vehicles. The lack of a need for the photon has been suggested by Heitler(1944), Bunge(1973), and Armstrong(1983) on similar grounds.

Nuclear Resonances: the Same Space but a Different Resonance

The two principles, the MAP and the Space Density Assumption can also provide an origin of nuclear forces. Consider what other kind of resonances could occur at the center of a charge resonance. If the center is a strongly non-linear region, it behaves like a transparent sphere with a high index of refraction that will guide light waves inside in a circulating mode. No experimental evidence indicates spherical symmetry of the proton nucleus, so space waves that circulate inside the sphere, like a doughnut, are OK. If the wave density is very large, the wavelength will be small and the velocity and frequency will be large, which means the mass will have a precise value since circulating waves are standing waves which must repeat themselves at a precise wavelength. Thus a proton can be composed of a + charge resonance with a high frequency nuclear resonance at its center.

This explanation is only a qualitative speculation. It would be much better to investigate nuclear forces by calculating possible resonant modes using the space resonance wave equation. This calculation is very difficult for a non-homogeneous medium. Unfortunately, I am a mediocre mathematician, but I hope that a clever mathematician will attempt this interesting but difficult challenge.

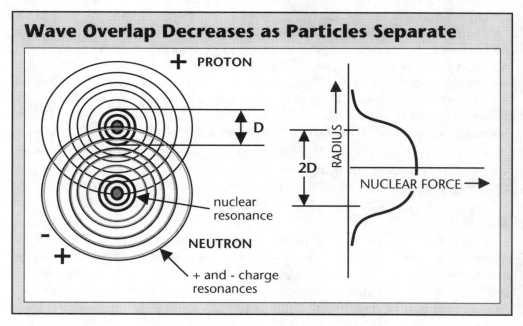

FIGURE 12-2. Forces between nucleons are short range.
The nuclear force is large when the central nuclear resonances overlap, but quickly falls almost to zero as they separate. Overlap is needed for the energy-transfer mechanism to transfer energy (change frequencies) between them.

CHAPTER 12/*The Space Resonance Theory*

Lacking calculation, I will compare this model further with measured nuclear forces. We already know many experimental facts about how neutrons and protons behave together. There are large forces attracting them together, especially for pairs; one neutron and one proton. A very tightly bound configuration is two pairs that form a helium nucleus He^4. The binding energy per He^4 nucleon is about 20 MeV, compared to about 6 MeV for the nucleon of an average atom. No reasons are known for this behavior although it is suspected that there is a connection with alignment of nucleon spins. How does this match the speculative nuclear space resonance?

Suppose two circulating nuclear resonances were superimposed, what would happen according to the MAP? The amplitudes would interfere either constructively or destructively depending on whether the circulation was the same or opposite. The MAP would cause strong attractive forces to retain the destructive configuration. So we can speculate that two aligned doughnuts can do this if circulating in the opposite directions. Conversely, if one doughnut was flipped, there would be no cancelling of amplitudes and force would be repulsive. (Remember, the two doughnuts have the same axis and occupy the same space). We deduce that only one neutron-proton pair can be together. So, a pair becomes a deuteron, and two of these are He^4. This agrees with experiment.

Consider how the forces change with distance. It is experimentally known that if two nucleons are separated by more than about e^2/mc^2, about 10^{-15} meter, the attractive forces quickly disappear. We note that this matches the distance already calculated, in equation (12-7) where non-linearity of space begins for space resonances. This is approximately the diameter of nuclear space resonances. So if the nuclear resonances are separated past this distance, their wave overlap falls off quickly, there is no energy transfer, and force approaches zero. So, again the resonance agrees with experiment.

The Unstable Particle Zoo. How does the nuclear space resonance explain the zoo of nuclear particles which have properties similar to a proton except that their masses are larger? All of these particles are known to decay quickly into a proton in 10^{-10} to 10^{-20} seconds. We suppose that the nuclear space resonance at the center of a charge resonance has higher modes of shorter wavelength, as is commonly found in other natural resonances. Thus the zoo is qualitatively understood.

We see that the proton is the mode of lowest possible frequency, hence it is stable. The more massive particles are, the higher the modes with higher frequencies and, depending on the mode, have different values of other parameters such as spin. They are unstable because there is a lower amplitude to which they can, decay, via the MAP, into a proton. The large density of waves at the resonance center guarantees a large coupling between resonance modes so the energy-exchanges are very quick, thus explaining the short lifetimes, typically 10^{-20} seconds, compared to atomic lifetimes of 10^{-8} seconds. Exclusion principles would be expected to exist because of orthogonal modes, thus explaining the slower decays of 10^{-10} seconds. All these interesting possibilities could be confirmed or de-firmed by solving the SR wave equation! Who will do it?

Another interesting problem with a valuable result is to see if a way can be found to match up nuclear space resonances with the group-theory explanation of the nuclear particle zoo. One of the names of that theory is the *Eight-fold way* discovered by Gell-mann and Ne'eman in 1960. It cleverly uses geometric groupings of the various particles to diagram their parameters: spin, parity, isotope number, and strangeness number. The group theory has not yet revealed a physical structure such as space resonances. If there is a relation it is logical to expect that solutions of the SR wave equation would have orthogonal properties that match the Eight-fold way. It is an exciting prospect to attempt.

There is a group-theory result about nuclear forces which results in a mathematical particle called a *gluon*. The gluon has an energy-transfer role between nucleons and other baryons like that of the photon between electrons. However, theorists are puzzled that a gluon has never been observed in any particle detector. If you interpret it using the space resonance idea, it is not puzzling, because the gluon, like the photon, is only modulation of the space waves traveling between the two nucleons. Clearly it makes no sense to expect this modulation to exist anywhere except between the two nucleons involved. Like the photon, gluons are not free to travel! Their only allowed destination is another baryon. No gluon detectors, please.

The Weak Nuclear Force: The Same Resonances but a Different Space

The *weak force* is a term which describes the very slow decay and split-up of some baryons into other particles. The most common example is the decay of a free neutron into a proton plus an electron, together with transfer of energy and momentum to other particles. The lifetime is about 14 minutes.

WEAK FORCE DECAY EQUATION: $n \rightarrow p^+ + e^- +$ **energy** (external transfer)

The key fact of the weak-force mechanism is that energy is transferred outside of the initial baryon. For this reason, the distance between exchanging nuclear resonances is large compared to nuclear distances. Again, we apply the the space-resonance energy transfer mechanism. We note the distance between particles in an exchange, from inside a nucleus to another particle outside the nucleus, is many times larger than the nuclear diameter e^2/mc^2. The nearest outside nucleon might be in an adjacent atomic lattice perhaps one micron

(10^{-6} m) away. This is about $10^{-6}/10^{-15} = 10^9$ nuclear diameters away. Thus the overlap of the wave amplitude of the central nuclear resonance with waves of particles outside is miniscule. Accordingly the coupling between them is tiny and the decay is expected to be extremely slow compared to the time it takes for an energy transfer times within one nucleus. Experimentally, typical decay times are 10^{22} times slower. For the same reasons, the amount of energy exchanged is small. The space resonance agrees qualitatively with observation.

An experiment to verify SR coupling. It is possible to experimentally verify the SR origin of weak forces because it predicts that decay times depend on the *coupling between two resonances* which may vary depending on the types of resonances and the distance between them. This contradicts the classical concept of weak decay which assumes that decay times are fixed because they are an intrinsic property of the decaying particle alone. For example, the decay time of the neutron, about 14 minutes on Earth, should become much longer if the decaying neutrons were located in outer space, distant from any matter.

Tetrode's Speculation

In 1922, H. Tetrode made the remarkable proposal that a particle never emits radiation except to another particle, "The sun would not radiate if it were alone in space and no other bodies could absorb its radiation.... If, for example, I observed through my telescope yesterday evening a star which is 100 light years away, then not only did I know that the light which it allows to reach my eye was emitted 100 years ago, but also that the star or individual atoms of it knew already 100 years ago that I, who then did not even exist, would view it yesterday evening."

At first glance, this idea seems crazy because it appears to foretell the future. But, read more carefully, his strange words are merely the consequence of a postulate that radiation is always an exchange between two particles; in itself not a revolutionary concept. An exchange of energy between two oscillators in the laboratory is commonplace. Tetrode logically said it can continue to be commonplace even if the oscillators are 10 meters apart, or 1000 meters, or 100 light years apart. And why not? Is there a magic distance beyond which exchanges no longer take place?

Tetrode was of course speaking against the concept of independent existence of photons, a new and popular idea at that time. (It is also probable that he was exaggerating when he endowed ancient stars with knowledge of people's existence.)

Like myself, Tetrode deduced that *radiation is an exchange mechanism*, like the SR theory. However Tetrode's most remarkable insight was that each radiating atom had some foreknowledge of where its radiation would finally end up. Then and now, this idea is radical! However, he was also forced to that conclusion, because if you postulate an exchange mechanism, the symmetry requires that the two partners behave similarly, and each must have some knowledge of the other at the beginning of the exchange. In the SR theory, this knowledge is provided without difficulty by the IN waves, but Tetrode had no idea of space waves; he only knew it must be so. And he had the courage to say it!

His exchange idea was intuitive speculation followed by logical reasoning and it had enough substance to affect the thinking of later cosmologists such as Wheeler and Feynman (1945), and Lewis in 1926. The major objection to it was that it violated the philosophy of causality; causes always precede events. The ideas of causality are now in trouble because of the EPR proposal and the Bell experiments which *have* violated it! To reconcile causality, we need to agree upon a re-definition of the meaning of knowledge and information exchange. Since acquiring human knowledge always involves energy exchange, the concept of an *Energy Transfer Mechanism* is immediately involved. Figure 13-1 and the text will suggest a redefined concept of observeable and non-observeable communication.

The Next Chapter will look at more applications of the space resonance, explain a few more paradoxes, and examine further the relation between microphysics and cosmology.

Cosmology is fascinating, perhaps because it relates to the enormous universe which is the home of all of us, or simply because it is full of mystery. At the same time, it is very speculative, so get ready to disbelieve!

"You can always tell a pioneer by the arrows in his back."

— L. Ovshinsky

Chapter 13

Applications of the Space Resonance Theory

Gravity and the Mystery of Dirac's Numbers
Newton's Law and the Mystery of Mach's Principle
Let's Try to Explain Quantum Puzzles
Exchanging Energy Between Particles
A Short Tale of Dirac's Infinite Proton
Refitting Science with the Space Resonance, a Unification of Physics
The Unity of Forces
High Energy Physics is Big Business
Let's Take a Second Look at Length, Mass and Time
Mastery of the Cosmology Game
How Electric and Magnetic Laws are Included
   Within Space Resonances

Chapter 13
Applications of the Space Resonance Theory

This chapter investigates the interesting mysteries of cosmology originally posed by the famous natural philosophers Paul Dirac and Ernst Mach. The SR theory is used to interpret them and another sensational puzzle of quantum mechanics proposed by Einstein and recently placed in the limelight by John Bell. These explorations seem to lead to an understanding of the origin of forces and the conservation of energy. You, the reader, are asked to judge whether or not these new ideas are likely to be verified. Do the logical tests for new ideas reviewed in Chapter 6 confirm them, oppose them, or leave them in limbo?

Gravity and the Mystery of Dirac's Numbers

Chapter 11 introduced three puzzling numerical relations discovered by Nobel Laureate Paul Dirac. Now we use the SR theory to explain his numerical relation,

$$e^2/Gm_eM_p \approx cT/(e^2/mc^2) \approx 10^{40}$$

and also to find an origin of the unexplained force of gravity.

The expression on the left is not a mystery, it is just the ratio of the electric to gravitational forces between an electron and a proton, a measured value of **0.23 x 10³⁸**. These numbers have no theoretical origin, they are simply the result of experimental measurements.

On the right side, you can speculate that **cT** is the radius of the universe, since **c** is the speed of light and **T** is the reciprocal of Hubble's constant. The divisor is the classical radius of the electron, e^2/mc^2. Why this ratio of the smallest and the largest entities in Nature should also be approximately 10^{40} is the mystery. Dirac found other ratios of size 10^{40} too. I will try to explain them by finding a theoretical value for gravity force using the SR particle model.

Since gravity is such a tiny fraction of the electric force, you may suspect that gravity is a perturbation or disturbance, of the electric force. A perturbation could be caused by the Hubble expansion of space which takes place very slowly producing tiny effects.

The IN/OUT waves which are the media of the electric force have been assumed identical, but this can't be precisely true if space everywhere is expanding, because the IN waves precede the OUT waves in time. The Hubble expansion will create a tiny wavelength change of the OUT wave compared to the IN wave, so that they are slightly different as they pass each other through the center. The IN waves of *other* space resonances will notice this difference and the *Minimum Amplitude Principle* will seek frequency changes and relative movement to correct the imbalance. The frequency change and movement will be observed by us as gravity forces.

Let's check this possibility by estimating the force change. The fractional change of space S (length) during a time delay Δt, is its product with the Hubble constant, or $\Delta S/S = H\Delta t$. This ratio is also equal to the force ratio $\Delta F/F$. The time delay Δt is equal to the distance traveled divided by the speed of the waves, or c. Approximate the difference of travel distance of the IN/OUT waves as the distance from center to the first node which is half the Compton wavelength or half of h/mc.

We have hypothesized that **F** is the electric force and ΔF is the gravity force, so we can write

electric force/gravity force $= F/\Delta F = 1/H\Delta t = cT/(h/2mc)$

Inserting known values of the constants (see table 2-2 in Chapter 2) the ratio can be estimated:

electric force/ gravity force = 0.12×10^{39}

This is near the measured value 2.3×10^{39}. This agreement involving such large numbers is remarkable considering the crudity of **H** and the approximation involved in estimating the time delay.

The expression $cT/(e^2/mc^2)$ on the right of Dirac's equation can be given the same meaning, and yields a similar result since h/mc (the Compton electron radius) and e^2/mc^2 (the classical electron radius) differ only by a constant factor 1/137.

We have found a meaning for Dirac's numerical ratios and a possible origin of gravity force. This provides some reasons for believing that the SR theory may be correct.

Newton's Law and the Mystery of Mach's Principle

Chapter 11 described Ernst Mach's startling proposal that Newton's Law of Inertia depends upon the existence of the distant stars. Mach observed that our reference frame for rotary motion appears to be determined by the distant fixed stars and proposed that the forces of rotational inertia were caused by the matter of the entire universe. Einstein agreed and thought he had incorporated Mach's principle into General Relativity but later decided it was not there. Narlikar (1984) also makes use of it, but there has been no proof that Mach was correct.

We will use the space resonance concept of radiation to examine the possibility that inertial forces are a tiny perturbation of electrical forces that result from acceleration of charge. When a space resonance (a particle) is accelerated, its arriving IN wave and the already departed OUT wave are not of the same wavelength as waves at the particle center, because of the Doppler shift. The differences will produce a shift in the total amplitudes of all waves in the environment of the particle. This will cause the *Minimum Amplitude Principle* to seek to correct this imbalance by readjusting the frequencies of the waves in the neighborhood. If the acceleration continues at a constant rate, the imbalance will continue at a constant rate and the rate of frequency (energy) change will be constant. Since the rate of energy change is constant the observed force is constant. This force is the same as the inertial force of Newton's law, F = ma. But is it the right size?

Notice that most of the waves in the environment of the particle are the billions of waves from other distant particles in the universe. When the *Minimum Amplitude Principle* adds all the waves together, these other waves dominate the total even though any one of them is infinitesimally small. They are, so to speak, a background wave amplitude of the particle under consideration. In view of their relatively large total amplitude, we conclude that these other waves would determine the degree of frequency shift and thus the force of inertia, including the numerical ratio between acceleration (meters/second per second) and force(newtons) of the MKS system of units. This is exactly what Mach had proposed.

Note also that as the wave frequency of the particle is increased by the MAP, its mass must likewise increase since mass and frequency are the same. This agrees with the calculation of the frequency change due to motion of Chapter 12 and the Appendix which matches the required relativistic increase of mass.

How can we find a calculation to verify the perturbation idea? If inertia is a perturbation of the accelerated charge forces, we expect that a calculation using mass instead of charge but otherwise following the same procedure as the electrical force should give a proportional result. It is a good idea to try, but we are not certain it works. For one thing, acceleration of a neutral atom involves both a positive and a negative particle, and we have to assume that both particles behave in the same way. The SR theory encourages us that they will be the same, since both sets of IN and OUT waves will be Doppler shifted in the same way by the motion.

Let us compute this analogy to see if it yields the right numbers. The electric field of an *accelerated* particle is a result of the relativistic change commonly termed "magnetic" (It is not the ordinary Coulomb force). If you read textbooks on electromagnetism, you find this electric field E is given by the time rate of change of the magnetic vector potential A; that is $E = \partial A/\partial t$. Using this we get

electric field = acceleration x e [$1/(4\pi e_o$)]/(c^2 x r)

For the analogous mass calculation, write

"mass field" =   acceleration x m [G] /(c^2 x r)

where charge **e** is replaced by mass **m**, the electric constant $1/(4\pi e_o)$ is replaced by the gravity constant **G**, and **r** is the distance to the charges or masses affected by the field.

To find the force of acceleration on the particle it is necessary to choose the masses which are influenced by the field. Since we are checking out Mach's idea, we have to choose all the other masses in the universe, M_u which is equal to the universe's mass density ρ_u times the volume of the universe or $M_u = \rho_u\, 4/3\pi\, (c/H)^3$. The Hubble distance c/H is taken as the radius of the universe. We assume the average distance of M_u is half the Hubble distance or r= c/2H. Then the force between the particle and M_u masses becomes

Force = Mass field x M_u = acceleration x m {($8\pi/3$) $G\rho_u$ /H^2}

To evaluate this force, use the present measured values of $G = 6.67 \times 10^{-11}$, $1/H = 4.72 \times 10^{17}$ seconds, and $\rho_u = 5 \times 10^{-29}$ kg/m^3. We find that the quantity in braces {...} is a *dimensionless* number $\approx 1/12$, very close to unity. If it were unity, and it could easily be so in view of the rough knowledge of H and ρ_u, then the remaining equation surprisingly becomes Newton's Law of inertia: **F = ma**! Note further that if we use the *critical density* of the universe of 6×10^{-28} kg/m^3 (according to general relativity this value is needed for a flat universe) then the factor in braces becomes 0.8, even closer to one.

Since the idea appears to work out correctly, we should review it to better understand it, and look for flaws. First, we have assumed that inertial forces are a small perturbation, analogous and proportional to the electric forces of acceleration which are due to special relativity (see Chapter 10 on magnetism). We have assumed that both positive or negative charges behave alike under acceleration, based upon the SR concept. Finally, we have assumed that Mach's principle is correct, so that all the masses of the universe receive wave disturbances from any accelerated mass. As a result, we obtain Newton's Law. You must think this through and decide if it is correct, probable, possible, or crazy!

Dirac's Second Mystery. If the above is correct, there is an another bonus (or you might regard it as an argument in favor of Mach's Principle.) Compare the quantity in braces {...} above with the second of Dirac's mysterious large number ratios (Chapter 11, p.174) which has laid unsolved for the last fifty years. They are identical! If you accept Mach's principle and the SR theory, Dirac's second puzzle is solved!

Let's Try to Explain Quantum Puzzles

Because the space resonances directly yield the deBroglie and Compton wavelengths as well as the $\psi^*\psi$ probability concept, which are the experimental foundations of quantum theory, the resonance theory underlies and supports quantum theory. The space resonances are also a QM physical model (formerly unavailable) and provide a long-sought basis for quantum theory. This physical model is now able to explain disturbing quantum paradoxes, such as the mathematical infinities due to point charges, and the strange result termed the "EPR paradox" (Einstein-Podolsky-Rosen), in Chapter 9 which appears to allow communication at infinite speed. The explanations of these two paradoxes by the SR theory will be discussed.

Bell's Theorem – Einstein and Quantum in Conflict. Chapter 9 described the story of EPR and Bell's Theorem, the strangest puzzle ever to come out of quantum mechanics. We summarize the story now, which you may wish to reread. In 1935, Einstein was philosophically opposed to some of the QM ideas and he used the most certain fact he knew, the impossibility of communication faster than light, to attempt to disprove a prediction of QM. This was the purpose of the proposed EPR experiment.

When actual EPR experiments were undertaken in 1972 through 1982 (Aspect, Dailibard and Rogers, 1982) the amazing result was that Einstein appeared to be wrong! In some mysterious quantum way, communication did appear to take place faster than light between the two detectors of the apparatus. These results show that our understanding of the physical world is profoundly deficient.

The proposed space resonance, particularly the behavior of the IN and OUT waves, is able to resolve this puzzle so that the appearance of instant communication is understood and yet neither Einstein nor QM need be wrong. In order to show this, it is necessary to carefully look at the detailed process of exchanging energy between two atoms, by the action of the IN/OUT waves of both atoms.

Exchanging Energy Between Particles

Perhaps the most fundamental process in Nature is the exchange of energy. It is the basis of all observation and measurement. It underlies our concept of force which we define as the *rate of change of energy with distance,* or dE/dx. Familiar examples of energy exchanges include the work machines do for us, the conversion of sunlight into food by plants, and the stimulation of our five senses such as sight and hearing.

When two space resonances exchange energy, there is a change of frequency of their oscillations. One oscillator frequency will increase and the other will decrease. Or, one resonance delivers energy and the other receives energy, which are the same thing. The equal and opposite changes underlie the law of *conservation of energy*.

Forces and Energy Exchanges are a Result of the Interaction of the Waves. Let's review the three requirements for a SR exchange to take place. Recall that there are always two resonances involved; one increases frequency, the other decreases:

First, the waves of each space resonance must *overlap* the non-linear wave-center of the other. Without this, there can be no modulation of each other's waves. This requirement is always met by charged particles whose waves extend to infinity, but not always in the case of nuclear waves, which do not propagate significantly beyond the central region of a resonance.

Second, the space at the wave-center must be in a *non-linear* condition so that it can modulate waves which pass through it, otherwise there can be no exchange of frequencies. This condition occurs in any resonance whose space at the center has been deformed by high wave density.

The **third** requirement, a property of space, is expressed in the *Minimum Amplitude Principle*, which states that two or more interacting resonances will change their frequencies and positions in such as way as to achieve a minimum total amplitude.

Analysis of this SR energy exchange process, below, will enable us to more deeply understand the unusual communication which occurs in the EPR experiments, and also the meaning of communication generally. As a bonus benefit, we will discover more relationships between electric, nuclear, and weak forces.

Communication can be Observable and Non-observable. Figure 13-1 illustrates the behavior of the IN and OUT waves during an energy exchange between two space resonances. It shows the series of events which take place as the IN and OUT waves communicate via their non-linear wave centers. Two communicating resonances behave much like two physical oscillators which are suddenly coupled together and readjust their oscillation frequencies. The IN and OUT waves play complementary roles in the transfer process. After the adjustment is completed, the receptor atom (resonance A) has a higher frequency and the source atom (resonance B) has a lower frequency. These final frequency changes are observable, equal and opposite in accordance with the conservation of energy.

There is a beginning, a transition, and final stage of communication. *Both* atoms are involved in these events because their continuing IN and OUT waves create the entire communication process. In a limited sense, both atoms know the wave state of the other atom at the outset, because the IN waves provide this limited information. Otherwise the exchange would not begin. But these precursor events are not observable because they are not marked by frequency shifts.

Communication is divided into two types. The *observable* type is the result of definite shifts of energy. That is, the "transfer" of energy from one atom to another. This type leads to the classical concept of the velocity of light (an observation of an energy shift). The *non-observable type* is associated with knowledge of initial state conditions of other atoms (resonances). No energy shift takes place.

The IN waves dominate the beginning stage by communicating the potential of decreased amplitude; if the initial wave states can be readjusted in accordance with the MAP. The arrival T_0, and modulation T_1, of IN waves are not frequency shifts and are not observable.

There are two observable events. The first $T_1 + \Delta T$, is when an observer at the location of resonance B would state that a "photon" had just departed. The second T_2, is when an observer at the location of resonance A would state that a "photon" had arrived. These two events are marked by a frequency change at the respective atoms and therefore are observable.

Step-by-Step Description of Figure 13-1: Sometime prior to time T_0, atom A had acquired an empty level (i.e., the detector is "turned on") and atom B is in an excited state which is resonant with atom A (the source is turned on). Then the following sequence of events begins with all waves traveling at velocity c.

TIME T_0. Part of the IN-wave of atom B, passing the center of atom A, responds to the *Minimum Amplitude Principle*, by seeking a lower minimum in combination with A, and begins a resonant exchange. This event is a precursor which begins modulation of particle waves and is not directly detectable. It is partly responsible for the mysterious communication in the EPR-Bell experiments. A similar sequence is followed by the waves of A.

TIME $T_0 + \Delta T$. The OUT wave of A begins to change frequency and the A-center modulates the IN wave of B in a mutual coupling of the two oscillators.

TIME T_1. The OUT Wave of A has reached the center of B which can now join in a mutual coupled exchange oscillation. If two polarized states are

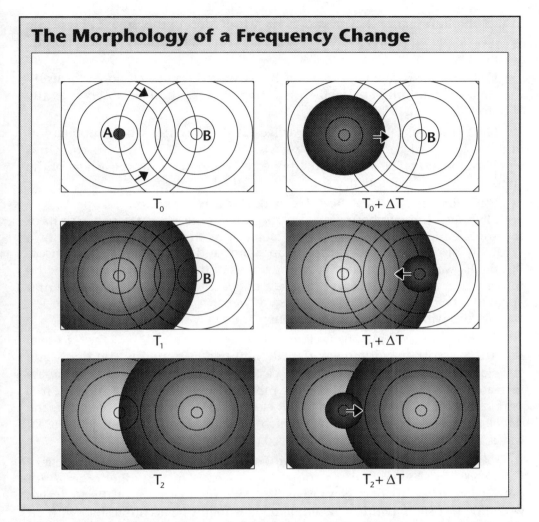

Figure 13-1. The Morphology of a frequency change (energy exchange) and the communication process.
   The IN and OUT waves of two atoms, A and B, carry information about the state of the other atom, behaving like coupled oscillators, eventually terminating in a readjustment of the two frequencies to obtain a lower total wave amplitude in the surrounding space. At the end, atom A (the detector) will shift frequency upward, and B (the source), will shift frequency equally downward, complying with conservation of energy.

involved, the fact that the waves find themselves able to respond to the MAP, is equivalent to atom B knowing the polarized state of A, relative to itself. To the observer of the EPR experiment this appears to be instant communication, but in fact the information has passed from A to B, during the time T_0-T_1, at the velocity c. The frequency of atom B begins to shift downward (i.e. a "photon" is on its way.)

TIME T_1 + ΔT. The OUT wave of B has shifted frequency down to nearly the final equilibrium value, which is interpreted by the EPR observer as the "photon" having left B.

TIME T_2. The OUT wave of B arrives at the A-center where modulation takes place, and the frequency of A shifts upward, nearly to the final equilibrium value. The event is interpreted by the observer as the arrival of the "photon," since the frequency shift is a measurable change.

Explaining the EPR-Bell Instant Communication. The passage of both IN waves through both wave-centers precedes the actual frequency shifts of the source and detector. A means to detect this first passage event is not a capability of the usual photo-detector apparatus and remains totally unnoticed. But the IN waves, are symmetrical counterparts of the OUT waves and carry the information of their polarization state between parts of the experimental apparatus before the OUT waves cause a "departing photon" event. The IN-waves travel with the speed of light so there is no violation of relativity.

At this point you may be inclined to disbelieve the reality of the IN wave. But there is other evidence for it. Remember, it explains the deBroglie wavelength and thereby QM. It is necessary to explain the relativistic mass increase of a moving object and the symmetry in its direction of motion. It is responsible for the finite force of the SR electron at its center. Especially, it is the *combination* of IN and OUT waves which explains these laws, not just the IN waves. If you believe in one you are forced to believe in the other.

Can Proof of the IN Waves be Found? Before one can really believe any theory, an experiment to show the existence of new phenomena not previously known is most persuasive. To prove the existence of the IN waves would be just such a critical experiment. This might be accomplished with an apparatus of the type used by Aspect, Dalibard, and Rogers (1982) except that instead of making a random filter setting *during* a "photon's" passage time, the filter setting should occur during the time period *preceding photon departure*. That is during period T_0-T_1 of Figure 13-1. The purpose is to frustrate communication by the IN waves, if they exist. If the IN waves are necessary to the energy exchange process, and if they really exist, then the result of the experiment would be a linear relation between the angular difference of the two filters and the number of coincidences. This would be the result originally expected by Einstein for the EPR experiment.

A Short Tale of Dirac's Infinite Electron

Despite the engineering utility of Coulomb's force equation, it has a history of severe theoretical problems mostly concerned with the infinite force, mass, or energy when radial distance to an electron approaches zero.

An interesting story from the history of QED provides a realistic description of how theoretical problems are sometimes solved. In the 1930's, mathematical theorists were wrestling with the structural problem of an electron which was imagined to be something like a sphere layered with charge substance. They wanted to account for the forces between different parts of the charge layer. If the electron was moving steadily, the forces would balance out, but if it was accelerated, they would not balance. Using Maxwell's Equations for electromagnetic waves from the accelerated electron, the theorists worked out an expression for the forces as follows:

$$\text{Force} = [k_1 e^2/(Rc^2)] x''] - [2/3] x''' + [k_2 R e^2/c^4] x'''' + \quad (13\text{-}1)$$

x'' is acceleration, x''' is rate of change of acceleration, x'''' is the rate of change of that...etc. R is the radius of the electron sphere, k_1 and k_2 are constants. Note the minus sign before the second term.

Since the first term alone is like Newton's law F = ma, the coefficient of x'' was called the "electromagnetic mass." Obviously, if **R** becomes zero, the electromagnetic mass becomes infinite. Since infinite forces are never seen, **R** could not be zero. But if the electron was an elementary particle, **R** had to be zero to avoid explaining what the sphere was made of. Further, relativity demands R = 0 to avoid communication problems between parts of the sphere. This was a dilemma.

Nobody wanted the first term and its troubles. The last and later terms didn't matter because the $1/c^4$ factor makes them too small to be observed. The second term with x''' corresponded to the energy of radiation from an accelerated charge, an observed fact. So a lot of ideas were tried in order to retain it. All of the ideas led to conflict with observation, until Paul Dirac said, "Look, we found the force equation (13-1) by using an *out-going* electromagnetic wave. Now if we assume a symmetrical but in-coming wave, the sign of the x''' term is postive instead of negative. Therefore we can make a new rule: *An electron acts upon itself according to one-half of the difference of the outgoing and incoming waves which it produces.*" This will cancel other terms and leave the one we want." No theoretical reason was found for this arbitrary procedure or to justify the in-coming electromagnetic wave. It was used just because it worked!

Dirac's scheme worked well for electromagnetic theory: The electromagnetic mass became zero and the radiation energy was right. But later when electromagnetism and quantum theory were joined in quantum electrodynamics (QED), the infinite mass reappeared because of the point electron. Once again, an arbitrary procedure was invoked. This was to cutoff the electron's field at a chosen distance to get the result they needed. In technical jargon, the procedure was called "renormalization."

Although we know now that the electron is not a layered sphere, many of the ideas from this period 40 years ago are still alive. Which ideas are still valid and should be retained? Which are wrong and should be corrected? Science has no automatic means of self-correction. Only new scientists with newer ideas can do this.

QED Discards Infinity. The electron of quantum electrodynamics (QED) cuts off the electron's $1/r^2$ force at a distance of about $r \approx e^2/mc^2$ in order to remove the infinate force at $r = 0$. The justification is that it works. The accuracy obtained using the procedure is so amazing it is hard to deny its correctness, but there is no justification for the procedure.

In the space resonance there is a natural cutoff because the standing resonance waves have a finite amplitude at the center. Because of this, the force between two SR electrons reaches a maximum at a separation about $r \approx e^2/mc^2$. It does not become infinite at $r = 0$. This unique feature of the space resonance explains why the QED cutoff distance is $r \approx e^2/mc^2$.

Refitting Science with the Space Resonance

Let us suppose that space resonances were the fundamental form of matter. Then, how would professors explain the behavior of particles and their interactions using this new model of a particle? Everyday physics and chemistry would be combed over and new answers provided to explain, "How are the atoms built?" and "Where do the rules come from?" In the following paragraphs, I will describe some of the basic interactions and particle behavior according to a future space resonance model.

These explanations will be less complex than they are now because the space resonance is a single building block replacing many previous ideas. It has a mechanism to exchange energy which was not known before. The formerly mysterious conservation laws will be understood and form a logical picture. There are fewer assumptions than before. Everything is easier to understand and remember.

The SR Minimum Amplitude Principle is the Master Mind of Space. In the future, lecturers may be explaining that: "The MAP principle controls all energy exchanges and thus every activity in the universe. Even though MAP is the great dictator, it is also benign because its only goal is to minimize wave amplitudes and thus smooth out the universe." It seeks calm rather than chaotic violence.

Every atom, comprised of different kinds of resonances, has a great variety of allowed oscillating modes; the more complicated the structure, the more are the oscillating modes. A complex atom such as iron, has thousands of commonly observed modes. Every mode corresponds to a different physical arrangement of the electron waves in the space around the protons in the nucleus at the center. The waves arrange themselves to satisfy the *Minimum Amplitude Principle* (MAP) and smooth the space occupied by the atom.

It is always possible to disturb an atom by introducing other particles and/or energy that cause the atom or molecule to choose a new arrangement. When this happens, the frequency, and thus the energy, of the wave motion will change. This change cannot happen to an atom spontaneously — there must be two atoms, one of which is decreasing frequency and the other increasing. The final combined arrangement of both atoms must be a net decrease of total wave amplitude, to satisfy the MAP.

An energy exchange only occurs when the frequencies of two oscillators match each other. This can happen if the two atoms involved are the same kind of atom. It could also happen if different atoms have identical level differences. Or, if the levels are not the same, each of their waves arriving at the other have the same frequency because of relative motion and the Doppler effect. This change of frequency of waves due to relative motion and the Doppler effect corresponds to relative kinetic energy.

Molecules as well as Atoms are Resonant Wave Structures in Space. An electron is the simplest space resonance, but combinations of electrons, protons, and neutrons to form atoms are also space resonances with the different particles arranged so that total wave amplitude is as small as possible, following the dictates of the *Minimum Amplitude Principle*.

Look at an example: The hydrogen atom, which has one electron and one proton, is a combination of three space resonances. The simplest H mode has spherically arranged waves with the proton as its center. The proton is two SRs superimposed upon one another; a nuclear resonance and a + charge resonance. Add a - charge resonance (the electron); then a neutral hydrogen atom results with the positive and negative resonances cancelling each others waves.

It is well known that the electron's wave structure can exist in a variety of resonant modes around the proton. When undisturbed, the atom chooses to remain in the resonant mode with the lowest energy, according to the MAP. The normal lowest energy mode has complete spherical symmetry. The next higher mode has two lobes around the center. Some of these are pictured for the H atom in Figure 9-1. If transitions take place between modes, the resulting series of energy exchanges to an outside detector are given names: the Balmer Series, the Lyman Series, etc, after their discoverers.

Spin, Spherical Rotation and the H atom. Angular momentum and spherical rotation, discussed in Chapter 5, are enigmas of the quantum world which tantalizingly beckon to us and suggest that they contain answers to the puzzle of particle structure.

The Classical Enigmas of Spin. Angular momentum in an atom is something like human-realm angular momentum and has the same units: mvr = kg x meter^2/seconds, but it also has strange features. Especially, we don't know what is rotating. The simple idea of a rotating particle mass doesn't fit. More perplexing is that the electron and proton have spherically symmetrical charge forces, so it is hard to see why they combine into an atom in an unsymmetrical fashion as some classic particle models require. There is also the knotty issue of the continuity of space: Any atom model which involves particles in orbital motion around a nucleus, must violate the continuity of space. A separate problem, not involved with the atom, is that electron spin seems to require an axis of rotation, which is clearly *not* spherically symmetrical. We want to examine these problems by looking at the simplest possible atom, the H atom.

It is shown in Figure 9-1 that the wave functions of the H atom have quantized solutions depending on quantum integers, L, M, and N. These integers have analogs in the behavior of standing waves on strings, in pipes and musical instruments. In Chapter 8 we saw that the number of nodes on a string was related to an integer n, which appeared in the sinusoidal standing wave equations. In the H atom wave function, we found exactly the same phenomena except that the spherical geometry makes the wave shapes more complicated than those which appear on a simple string. Additionally, the H atom is 3D so there are three quantum integers: L, M, N which regulate the 3D standing waves. Two of these, M and L, are related to angular variables ø and Ω which rotate. From Figure 9-1 we observe that the quantum wave patterns are standing waves around the circumference of the sphere of spherical coordinates. How do they arise?

Explaining the H Atom Quantum Wave Patterns. Spherical rotation (Chapter 5) and the SR model have not yet been applied to the hydrogen atom. Let us apply it by looking at the relative motion of the proton and electron in the H atom and see what we find. What motion should we examine? We are dealing with QM waves so we must examine the relative velocity between the electron resonance and the proton resonance. This relative velocity produces a QM wave from the Doppler shifted quantum wave of the electron which arrives at the proton and vice-versa.

The wave patterns of the QM equations have been plotted in Figure 9-1. The three upper-left patterns are the normal lowest energy state of the H atom, and are completely spherically symmetrical. These are what would be expected from the IN/OUT waves of the SR electron. Consider how the remaining patterns arise.

Azimuthal Quantum Integer, M: The quantum integer, M, labels standing waves which circulate around the z axis. These standing waves arise from *spherical rotation* of the electron resonance with respect to the proton resonance. When M = 1, there is one wavelength in a circle around the axis. When M = 2 there are two waves, etc. After every two spherical rotations, the nearby space returns to its initial configuration. Thus there is one wave cycle in the space for each two spherical rotations. The only allowed rotational velocities are those which produce an integral number of wavelengths around the 2π circle of rotation, otherwise the standing waves cannot exist. This requirement reproduces energy levels which match experimental observation just as do other theories.

The spherical rotation does not result in any changes of frequency and so there is no energy associated with the motion. The effects of rotation are purely quantum mechanical and show up only as the pattern of the waves. By postulating spherical rotation, it becomes possible to have a SR particle structured out of space, with rotation that does not destroy the continuity of space.

Notice in Fig 9-1C that the QM wave function is: Amplitude = $A_o \exp\{iM\phi\}$, which is a complex number, $\cos\phi + i \sin\phi$. Why is this? It is because the mathematics of 3D space are not capable of representing spherical rotation. Instead, four variables are required. We saw in Chapter 5 that four variables of a 4D hypersphere can represent spherical rotation. Don't conclude that there are four dimensions in our world! No, it is just a mathematical method needing four variables. The complex number supplies the extra one.

The spherical rotation leads to standing waves in roughly a spherical shell, close to the center. But far away from the center, the electron's IN/OUT waves are negligibly disturbed; they preserve their spherical properties unaffected by the rotation near the center. This is exactly what one needs to explain the electron's properties, but how does it happen? It is the unusual character of spherical rotation that space is deformed cyclically during rotation but returns to an undeformed condition every two turns. Disturbances at the center can only be propagated outwards at the finite velocity of light, and are thus attenuated.

Polar Angle (Ω) Quantum Integer, L: Rotation is also possible around axes perpendicular to the z axis of the M integers. This occurs in Figure 9-1D, for the quantum integer L which numbers the standing waves which travel between the poles of the z axis. These waves must also obey the rule that standing waves reinforce each other every 2π degrees of rotation, and so they are the same or a multiple of the M integer wavelengths. For this reason, the formula for the L integer wave function is a function of M as well as L and $\cos \Omega$.

When L = 0, there are no standing waves which is the normal atomic state. If L = 1, there is one wavelength between poles, if L = 2, there are two waves, etc.

When studying Figure 9-1, it should be recognized that the azimuth waves, the polar waves, and the radial standing waves are each factors in the total wave function, which is a product of all three. For clarity, the figures illustrate the three kinds separately and only one quantum integer at a time is varied. If I attempted to draw all three kinds of waves at the same time mixed, for example with N = 1, M = 3 and L = 4, then the standing wave shapes would be hopelessly complicated and difficult to see. Nevertheless, the whole wave function must be considered for completeness because it determines how quantum realm behavior gradually turns into human realm behavior as the number of quantum events, and/or the size of quantum integers are increased.

Radial Quantum Integer N: Figure 9-1 shows standing waves in a radial direction for varied radial quantum numbers n = 1,2,3. The standing waves are unusual because the IN and OUT waves are traveling between two points of reflection, one at the center where r = 0 and the other at infinity where r = ∞. Needless to say, this is an uncommon configuration. But since a radial standing wave between a small sphere and a large sphere, say r = 1 and r = 10, should surprise no one, there should be no objection in principle for the spheres to have radii of $r = 10^{-12}$ and r = ∞. You can look upon the r = ∞ sphere as being effectively much closer since the radial wave amplitudes quickly decrease because of the 1/r effect of 3D space.

Notice that in the SR wave model, the proton and the electron resonances are concentric. This means that their wave centers are identical and the IN/OUT waves travel in the same space without interference. This is typical behavior for waves, produces no philosophical objections, and agrees with calculated QM wave functions as in Figure 9-1. However, in a theory using "particle substances" or orbits, there is great objection to two objects occupying the same space at the same time.

Proton and Electron Charge Resonance Waves Must be Identical. If we accept the above assumption that standing waves are established in the space between and near the proton and the electron as a result of the combination of their SR waves, then we have implicitly assumed that both those sets of waves have the *same* wavelength and frequency. Otherwise, they could not combine to form standing waves. This leads to another far-reaching conclusion: *All charge resonance waves are identical, whether originating from electrons, proton, neutrons, pions, etc...* This strongly suggests that charge resonance waves are a property of space rather the particles, so that there is only one type of charge wave propagated by the media of space. Is this true? We can't be sure, but it appears logical that this concept, which I call *Charge Wave Identity*, and the SR concept are both correct, or both wrong.

Why is the Speed of Light Always Constant? Dozens of experimenters have measured the speed of light and they always have obtained the same answer c, regardless of the speed of the source relative to themselves and their detector. I have read explanations of this result many times and I confess that I have never understood it. My mind demands a cause and the standard explanations don't provide one.

But using the space resonance energy transfer mechanism, it becomes quite clear why experimenters always obtain the same velocity. To see why, you must first set aside the notion of a photon traveling independently in space, and replace it with the modulated IN wave of the receptor atom which delivers the information of the frequency change (energy exchange). The receptor atom is always part of the observer's detector apparatus, and therefore its IN wave is always in *his* frame of reference. The velocity of the IN wave is always c with respect to its own atom (the observer's detector). The experiment, therefore, isn't involved with the relative velocity of the source. The different observers are always measuring the same thing, their *own* IN waves. It is not surprising that they always get the same result.

These arguments seem logical to me and I hope they turn out to be correct. There may be more subtle reasons for the constancy of velocity c, but until we understand more about space waves, I am willing to accept that it is due to the IN waves.

The Relativistic Mass Increase has a Cause. The space resonance also makes it easy to understand relativistic increase of mass (frequency). Relative motion causes a Doppler induced frequency shift of the resonance waves. Since mass and frequency are the same except for conversion factors, this is seen as a mass increase, too. No problem. The fact that mass change does not depend on whether motion is to or from the observer also comes easily out of the algebra of summing two symmetrical Doppler shifts. That is all there is to it. The enigma is gone.

It is important to notice that both the mass change, and the symmetry with respect to direction of motion are necessarily dependent on the existence of *both* the IN and the OUT waves. They work as a team.

The Unity of Forces

The space resonance energy-transfer mechanism embodied in the *Minimum Amplitude Principle* offers an explanation of the forces of Nature: electric, gravity, nuclear, and weak forces. These natural forces are unified because of their common origin in the MAP which seeks a minimum total wave amplitude in space for all arrangements of space resonances.

Nuclear Force. This short-range force is associated with the neutron and proton. Neither its cause or its exact form are known, but there have been many measurements of the path deviations when one nuclear projectile passes near another nucleus. From these measurements it has been determined that:

1) The idea of a "force" to describe energy changes between nucleons is not a useful picture. What happens is better described by an "all or none" energy exchange because a smoothly changing dE/dr does not occur in nuclear experiments. Instead, the term "binding energy" is often used. The binding energy ΔE exchanged when a nucleon moves apart from others, has a fixed value in a given nucleus. This energy depends on how many nucleons are already in the nucleus. If you add or subtract one nucleon to a light nucleus like hydrogen, ΔE is about 20 MeV, whereas if you remove a nucleon from a heavy lead nucleus, a ΔE of only about 5 MeV is required.

2) Nuclear energy exchanges occur when the distance between nucleons is about 10^{-15} to 10^{-14} meter and does not occur at greater distances. It is probably not a coincidence that this is approximately the proton Compton wavelength **$h/m_p c = 1.3 \times 10^{-15}$ meter or the classical electron radius $e^2/mc^2 = 2.8 \times 10^{-15}$** meter. This is also the size of the non-linear region of the SR wave center, where the proton resonance is presumed to be.

3) The attractive character of nuclear energy exchange has only been measured for the proton and the neutron. There is no knowledge for anti-particles or a possible repulsive force, say for anti-protons versus protons.

4) Nucleons have electric charge of ±1 or zero, and charge force appears to be independent of the nuclear binding energy.

High Energy Physics is Big Business

Measurement of nuclear reactions between exotic particles in high-energy accelerators is a field of intense and expensive activity, and a wealth of data exists, detailing their energy levels and the rates of decay of the evanescent entities which are created, live for a microsecond, and then return to a placid life as protons. This information has been used for a variety of interesting theories concerning the structure of baryonic matter. The popular names of these theories are as exotic as the participants they describe, such as: "The eight-fold way," "Quantum Chromodynamics," "The Big-bang," "GUTS," "Quarks," and "Wormholes."

The names given to mathematical entities of the theories are decorations of the scientific stage, so that quarks come in colors of *red*, *green*, and *white* and have flavors of *up*, *down*, and *strange*, while energy is transferred between them by *gluons* and *glueballs*. These names and meanings have little to do with the actual mathematical properties they represent in the theories. Please note for future use however, that the names *vanilla, chocolate,* and *strawberry* are not yet claimed!

How can the SR Theory Describe a Nucleon? Space resonances, whether alone or joined with a nucleon, possess the property of charge and extend to infinity as $1/r^2$. They are the same resonance for protons and positrons, and similarly, an anti-resonance describes the electron and anti-protons. If the net charge is zero, as in a neutron, both a resonance and an anti-resonance (+ and -) are present. In this case, the waves neutralize each other, but do not annihilate because they are not identical at their centers, having in one case a proton resonance at the center, and being alone in the other.

The heavy proton mass is the result of a dense space, which occurs because the wavelength is 1/1840 of the electron. At this smaller radius we would expect that the wave density is $(1840)^2$ times larger. The frequency of the nucleon increases accordingly and the mass, is 1840 times the electron mass.

These qualitative explanations would be better calculated, but the wave pattern of the proton resonance is not easy to calculate. The waves must satisfy a wave equation, like

$$\nabla^2 \Phi - \{1/c(r,\theta,\phi)^2\} \partial^2\Phi/\partial t^2 = 0$$

which is the same as the *SR Assumption I* (Chapter 12) except that the velocity $c(r,\theta,\phi)$ is now a function of the space density or position in the resonance. Such an equation is very difficult to solve, nevertheless it may have solutions. I haven't found the equation of a proton resonance yet because it is like a famous remark about the the bumblebee: "Bumblebees cannot solve Stokes' equations, but somehow they manage to fly." You can imagine the proton resonance something like a light-wave trapped inside of a dense crystal ball. The waves circulate inside, forever curving because of the varying space density.

You can ask, "If the nucleon resonance waves are trapped, how do the charge SR waves manage to get in and out?" It is because the charge SR waves are spherical and concentric with the proton center. So they always enter and leave perpendicularly to the gradient of the central space density. Consequently they are slowed, but not trapped, during their passage from IN to OUT.

You may ask, "If gravity is a perturbation of the electric charge SR waves, how can the gravity mass of a proton be larger since only the charge SR waves travel to other nucleons?"

This is because the perturbation energy exchange takes place at the resonance centers where space is non-linear. According to the perturbation idea introduced in Chapter 12, the strength of the gravity perturbation is due to the time difference of IN and OUT waves *caused by the density at the resonance centers*. There is a large additional delay caused by the proton resonance, so the energy perturbation, or mass, is also larger. Thus this concept of the proton structure agrees qualitatively with the explanation of gravity forces.

The structure of the proton center resonance may be something like Figure 13-2.

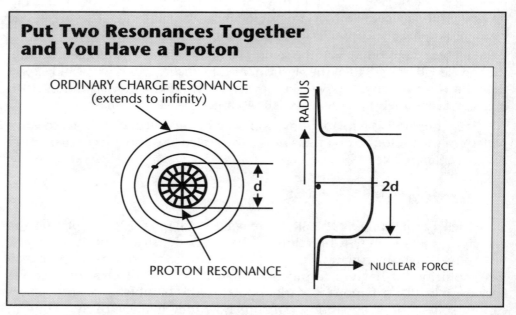

Figure 13-2. The hypothetical proton structure.
The proton may be an ordinary (charge) resonance combined with a dense "nuclear resonance" at the center. The nuclear resonance wavelength and diameter are small and intense so the srongly non-linear region of space traps the nuclear waves. The IN/OUT waves of the charge resonance are perpendicular to the density gradient and are not trapped.

According to the Energy Exchange rules, energy transfers may occur only if nuclear waves overlap, thus accounting for short-range nuclear forces.

The neutron has to be envisioned as a combination of a – charge SR, a + charge SR, and a proton central resonance. Obviously it has the same ingredients as a hydrogen atom, and if not confined to a nucleus, it can decay to an electron and a proton, forming an H atom after the excess energy and spin is transferred. The stability of the neutron, when combined with another nucleon, can be explained by the extra central density which does not allow the electron to find a lower MAP total amplitude by separating from the proton.

Let's Take a Second Look at Length, Mass and Time

In Chapter 2, we tried to understand the simple basics of science: *length, mass* and *time*. We knew that, in practical terms, these three elements of measurement were easy to define using international standards, but when we attempted to examine them minutely, their reality faded into the unknown realm of *space*. Can we understand them any better now?

Length, in our minds and in our laboratories, is a property of rods, tapes, and rules, etc. which we use for measuring. If we want to understand length we must think of what determines the structure of the materials of those rulers. From this viewpoint, length is determined by the electron waves in the lattice of molecules which form the material. In their reflections back and forth, the electron waves must have just one standing wavelength that fixes the spacing of the lattice nuclei. This wavelength is a function of **c, h, e** and **m** of the electrons which, in turn, are determined by the properties of space, especially space density. The evidence indicates that density depends on all other matter in the universe. Thus, length in the final logical analysis, like inertia, depends on the matter of the universe, as proposed long ago by Ernst Mach. We don't know whether the amount of matter during the history of cosmological times has been fixed or variable so we are not able to determine whether e, c, h, and m have been constant. We are ignorant because our history of science in our tiny solar system has existed for only a brief moment of time and we are just a speck in the space of the universe, beyond which we cannot see.

Mass, as we have observed and deduced, is a frequency whose value is obtained from: $hf = mc^2$, where m is the mass of an elementary particle. Our understanding of mass involves understanding the frequency of the elementary particles which make up all massive bodies. The SR concept implies that this frequency, in the case of electron and positrons, is that of the charge space resonance, and it obtains its value as a basic property of space. In the case of protons/neutrons, the nuclear frequency arises from the more dense non-linear space at the center of a SR. Again, its value is a basic property of space. Mass displays two faces: one, when an inertial situation is involved and Newton's law $F = ma$ applies, and two, in a gravitational situation where the law $F = Gm_1m_2/R^2$ applies. Accordingly, we need to understand the mechanism of both of these laws. The SR idea is alone in offering an explanation for these laws; namely that those laws result from a tiny perturbation of electric forces induced by the expansion of the universe. Presumably, if the Hubble expansion reversed itself, gravity would become repulsive and matter, indeed the solar system plus stars and galaxies, would fly apart!

Time in the SR theory has no new strange features. It is plain and simple. Those strange ideas which are raised about time in scientific speculations of other books; time travel, backward time, and 4th dimension time, and black-hole time are neither needed nor fit into the theory. The role which backward time plays in QED is replaced by the behavior of the IN waves of the space resonances, which travel inwards using the same time as the OUT waves. Both IN and OUT waves travel in the same normal time.

Nevertheless, time as a measurement in our laboratories and as shown on our clocks and watches also has an origin in space from the way space determines the properties of the basic particles (resonances). As the builders of "atomic clocks" already know, the measurement of time is ultimately derived from the internal oscillations of molecules, atoms, and particles. These large frequencies are too impractical to control any humanly useful clock. So the inventors of solid state computer chips have found many ways to mix large frequencies together, detect the difference, and then use that much smaller frequency to control a clock movement. So, we conclude that the basic clock of the universe is the frequency of the charge space resonance, a property of space.

We don't know whether time is alike in different parts of the universe, or whether or not it has always been the same during all periods of cosmological history. Like *length* and *mass,* our ignorance stems from our limited perspective of the universe. It is an interesting philosophical question whether we would be able to recognize a change of time-rate, or length, or mass, or any of the fundamental "natural constants." If it occurred, several of them could be changing simultaneously and maybe we wouldn't notice it. Since all measurements depend on these elements, we wouldn't know whether our instruments or the thing we were measuring had changed. We understand so little about space that, at present, we have no way of interpreting the possibility of varying space properties. A few good space-thinkers are needed!

Mastery of the Cosmology Game

Almost every philosophically inclined cosmologist has put forth his concept of how the universe began. It is an open game, with room for just about any reasonable idea that doesn't conflict terribly with fundamental laws. The reason for this freedom is a lack of experimental evidence to direct ideas into the correct direction. Such evidence is hard to obtain. The universe is a very big place and we don't know how to get out there and do experiments. Moreover, the universe has been around for a long time and few clues remain to tell us about the beginning. On the other hand, it is still a profitable game to come up with ideas, since a new cosmology can be the subject of a new paper and in a "publish or perish society," publishing papers counts!

However, you can't start a new idea which is completely groundless. There must be some glimpse of a principle or natural clue involved. If the principle has a catchy sound or a clever analogy that brings a chuckle, so much the better. Using the math of general relativity, Sir Arthur Eddington created the "dachshund" universe in which the universe's density rises, remains steady for a long time, then rises again just like the shape of the dog. Because the universe stays at one density for a long time (the back of the dog), it could explain why astronomers observe a lot of red shifted objects with a a shift near 2.2. Paul Dirac created a cosmology using one of his famous numerical coincidences (Chapter 11) in which G, the constant of gravity, and H, Hubble's constant, are related inversely. He views Hubble's constant as decreasing throughout cosmological time, therefore G must be increasing.

The presently popular idea is the "Big-Bang" (Chapter 11), but a "steady-state universe" has also been proposed in which nothing has ever changed. Several "oscillating universes" have been offered, some of which oscillate smoothly, contracting to a finite size and expanding again. Yet, another repeatedly collapses to zero and then rises out of the ashes like the legendary phoenix bird. Yet, another universe has time running forwards during one half of the cycle and running backwards during the other half. This is acceptable because nobody really understands what time is, so anyone who objects to backwards time is also treading on thin ice! Universes existing simultaneously have not been overlooked; two universes..., five..., a dozen, all are viable. Although the ideas began in the Sunday comic strips, it is only after mathematics are applied that they are recognized, on a "solid" foundation.

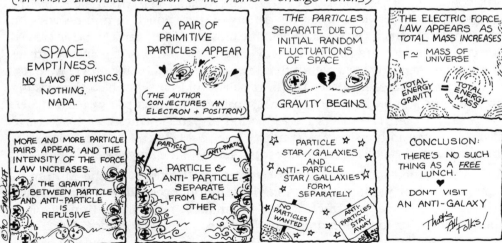

Figure 13-3. Creation of the Universe
The illustrator of this book began to think that all cosmologists were crazy and drew this version of the SR "kind and gentle" origin of matter.

Perhaps the favorite basis for cosmologies is the theory of general relativity. The advantages are legion. The name of Einstein associated with the theory adds no small amount of glamor and anyone who attacks your ideas can be counter-attacked by charges of desecrating his memory. Best of all, the government bureaucracies like to get on Einstein's bandwagon too, and are generous with money grants. Of course, since there is little proof that the theory is right or wrong, authors are free to roam, as long as they stay within the theory's mathematical boundaries.

A Kind and Gentle Origin of the Universe from Space Resonances. I, too, am not immune to the intrigue of the cosmology game. In addition to suggesting how the fundamental particles are formed out of space waves, there is a cosmological concept embedded in the SR theory. The basic assumptions of the theory imply how particles were formed, how the universe began, and what happened subsequently: Keep in mind though, that this "creation theory" is as as unprovable as all the rest.

In the beginning, all matter and activity was zero; no waves, no matter, no energy, and the physical constants were null except that *space existed* with a potential for propagating waves. At an initial moment of creation, two space resonances (a particle pair) formed and others have continued forming ever since. The reason for pair information is obscure — it can only be conjectured that a random fluctuation of space properties occurred.

Space resonances are always created as pairs: one particle, one anti-particle. The conditions of their formation is seen in *SR Assumption II* which states that the frequency of a wave depends on the density of space which in turn depends on the density of particles waves already in the universe. Thus, particles being created affect those which are already there. Going back to the first pair of particles formed, the frequency (energy) of those two was negligible since there were not yet any other waves in space. The next pair would have doubled the existing wave density and the frequency of the first pair, and the next pair would add 50% more, and so on. The successive formation of particles increases space wave density like the arithmetic series: density = $1 + 1/2 + 1/3 + 1/4 + $ ...etc. This results in a gradual and gentle increase of the total density of the universe.

As explained previously, electric forces are the result of the Minimum Amplitude Principle influencing individual resonances. Since a + resonance would be electrically attracted to a -resonance, re-combination of pairs immediately after formation must have been a common occurrence. Permanent matter formation only occurred whenever a + or - proton was available to join with a - or + electron to form stable hydrogen or anti-hydrogen.

The expansion of the universe also begins according to the M.A.P. which attempts to smooth space, by moving + or - particles with respect to each other. In the overall universe, the only general motion possible was separation of all resonances by expansion of the space of the universe. The Hubble constant describing this expansion rate decreases in value as the amount of matter and cosmological time increases.

Gravity began as a perturbation of the electric force because of the Hubble expansion. In order to avoid mutual annihilation of hydrogen with anti-hydrogen, gravity had to bring about separation of the hydrogen and anti-hydrogen. I deduce that matter and anti-matter are gravitationally repulsive. Matter must be clumped together in separate regions of the universe, and anti-matter in other regions. Therefore, 50% of the galaxies we observe must be of matter and 50% are of anti-matter. There is no way we can tell which is which since the radiation exchanged is identical for either type.

The physical constants are not constant, but continually change as the amount of matter in the universe increases. As matter increases, its affect on already existing matter becomes proportionately less, so that Hubble's constant H is becoming smaller. The constant of gravity G, caused by the Hubble expansion is also becoming smaller. Electric and quantum constants are similarly changing. Our laboratories observe them to be constant because the amount of change during human history is too small to be measured, but in cosmological times it has been large.

Is there evidence that this cosmology might be the right one? The evidence is mostly philosophical. This cosmology contains Mach's Principles and Tetrode's conjecture and, thereby, contributes to understanding of inertia, the photon, and energy exchange. It provides an explanation for the "missing" anti-matter which many persons think ought to be in the universe in equal amounts with matter. A reason is provided for the Hubble expansion. It leads to all the conservation rules for the universe: zero sums of charge, momentum, hadrons, and leptons, which have not been explained yet by any other means. It is also is in agreement with Edward Tryon's concept of a zero sum of energy and matter (Chapter 11). And, of course, it is not in conflict with the SR theory. Nevertheless, like other cosmologies, it hangs on mere threads of experimental verification. Which one do you favor, now?

Explaining double-slit experiments. One very frustrating paradox of quantum mechanics is the double-slit experiments described in Chapter 9 and Figure 9-4. A beam of particles is directed at two holes in a plate with a detector on the outgoing side to determine how the particles are distributed after they go through the holes. The experiment is done first with both holes open, then with only one hole open, and then with the other hole open. The particles distribute themselves on the detector side, as if they were guided by a wave; a quantum wave with frequency and momentum obeying $E = hf$ and $\lambda = h/p$.

Amazingly, the sum of particles which go through the holes open separately is not equal to the particles which go through both holes open together! Other apparently illogical results can also be found, using other thought experiments. These situations are called, "The double slit paradoxes."

As explained in Chapter 9, there are no paradoxes if you ignore the presence of the particles and only look at the waves. But, apparently, you can't ignore the particles because they seem to be necessary to transport mass and charge from one collision event to another.

But think, if the waves *alone* could transport those properties, there would be no paradoxes. This is one key element which allows the SR theory to explain the contradictions. The *space resonances* do transport these human realm properties, because charge and mass (frequency) are part and parcel of the waves. No particle is needed.

Energy or mass is carried as the frequency f, or more often as a change of frequency Δf of the resonance. Momentum is the change of wavelength $\Delta \lambda$ of the resonance. Charge forces arise out of the *Minimum Amplitude Principle* and the *Space Density Assumption* acting on the resonances. The property of "location" is a result of the two particles that are always involved in any exchange. One of these two is a *detector* and the locations of intervening detector-particles create the illusion of a moving "photon" acting as a carrier. The wave resonance model leads to no contradictions, and provides the large-scale human realm picture of the quantum world which we observe.

How Electric and Magnetic Laws are Included within Space Resonances

It is not necessary to demonstrate that each of the laws of electricity are encompassed by the proposed space resonances. We have already seen that space resonance properties include Coulomb's Law of the elementary charge and the laws of special relativity. Those two laws are a sufficient proof, because in Chapter 7 and the Appendix it is shown that these two alone account for all of electricity and magnetism. Therefore, if a space resonance possesses these two properties, all of E and M is encompassed. Still, it wouldn't be a good story if I didn't say something more about the important electrical waves, so I will describe some of the highlights below.

Electromagnetic Waves Travel Between Space Resonances. For a clear understanding of electromagnetic waves, you must first decide whether you are thinking about one wave traveling between just two atoms, or about a lot of waves traveling between a lot of transmitter and receptor atoms. If you mix them up it is easy to become confused because both situations are not the same. The primary reason for most of the confusion is the smallness of Plank's constant h, which means that the energy of a single wave $E = hf$ is very tiny. So tiny that, until the last half century, no one had ever noticed a single wave energy exchange.

In the case of TV, FM radio signals, and most visual communication, many single wave exchanges are involved. Usually these signals are being continuously transmitted. You do not observe the quantum effects that occur in the case of one wave. Instead you are seeing the *average* of many quantum events. Quantum events occur very quickly, much faster than light meters, TV receivers, or radios can respond. You don't see them, instead you observe that the signal appears to travel as a continuous wave between transmitter and receiver. The wavelength, frequency, and velocity c are all constant. From the continuous transmissions, interference patterns can be seen when the source is divided into two rays and brought together again. The patterns reinforce the illusion of a continuous wave in space. One never realizes that an enormous number of small events are creating a larger image. Just like all the dots in newspaper print create the image of a continuous photograph.

Low Frequency Energy Exchanges Are Radio Waves. Again, let us look at the behavior of a radio transmitter and receiver. Consider two identical resonances (electrons) which are part of copper atoms; one here in a receiver antenna and one somewhere else in a transmitter antenna. Each resonance is comprised of two continuous space waves: one wave is INward going and one is OUTward. If the centers are not moving with respect to each other, their waves pass by each other and nothing happens. Why? Because the waves are identical and there is no way to reduce their total amplitude according to the Minimum Amplitude Principle. Now, let one of them be forced into an oscillatory motion. The Doppler effect causes its OUT wave to become modulated from the view of the second atom. That atom readjusts its resonance frequency and motion in such a way as to minimize the wave amplitudes in the space around it. In short, it imitates the frequency and motion of the first atom. One has become a transmitter and the other a receiver. The IN and OUT waves are modulated and the modulation frequency is the observed FM, TV or radio frequency. If, as is usually the case, the radio signals are modulated with a news broadcast, there is modulation of the modulation of the space waves. The important point is that the actual traveling waves are the space IN and OUT waves, not their modulation, although that is the radio signal we observe. The space waves are not observed, we only see the energy exchange in the receptor atom, and we describe it as electromagnetic waves, which are actually the modulation of the space resonances.

An Example of a Single Photon Exchange. To understand a single energy exchange, it is necessary to review the energy-exchange mechanism in Chapter 12. Consider a sodium atom in a hot sodium lamp. Surrounding the nucleus are 23 space resonances (electrons); each one consists of an IN and an OUT wave. One of those electrons has been driven to a state of energy (frequency) higher than normal. Typically, the added frequency is the yellow light we obtain from a sodium lamp. That electron (resonance) can return to its normal frequency if a minimum combined amplitude can be achieved between it and some other resonance. Note that it cannot simply return by itself. It is in a higher state because it found a minimum condition there as a combination of its amplitude with the driving-cause amplitude. It will stay there until it finds another suitable combination of minimum total amplitude.

A suitable other resonance may exist in the pigment of your eye or in a phototransistor of a TV camera. An exchange may occur if the exchange-probability is sufficient. This will depend on relative wave intensities and environmental factors. If the exchange takes place the sodium electron will lose energy and our eye will gain energy to become part of a visual image of a sodium street lamp. We interpret the whole process as "photons" moving from lamp to eye.

In the next chapter, I will describe some possible ways in which this theory might be verified, and what might be ahead on the road to adventures in space.

"Beauty is Truth and Truth is Beauty."

— John Keats (1795-1821)

CHAPTER 14

Conclusions and Future Adventures

What Have We Found?
Can We Prove the SR Theory Using the Outcome of Experiments?
Predicted Theoretical Calculations Could Also Provide Evidence
Nature and the Universe Have Simplicity and Beauty
Where Can We Go From Here?

CHAPTER 14
Conclusions and Future Adventures

What Have We Found?

It is time to summarize the evidence found in this book's exploration of the universe, and to try to decide whether a wave model really represents the truth of particles. You may have made up your mind already, although I hope you will always be prepared to accept new evidence whenever it arrives. Now I will summarize my conclusions for you.

The Wave Model is Simple. Although the wave-model particle is very different from our human scale conception of a particle, it is quite simple. It is just a spherical oscillator located in space whose behavior depends upon the three assumptions of space properties: I Waves, II Space Density, and III Minimum Amplitude. The SR contrasts with the complicated classic electron which resorts to *ad hoc* photons in order to explain energy exchanges, and assumes charge and mass substances to ease our emotions. Electrons create "particle seas," "vacuum," "photons," and "bare mass," for which there is little evidence. Even so, the classic electron leaves dozen of questions unanswered, and a score of enigmas to confuse us. In contrast, the SR wave model is simple and neat, but is it right?

Space Underlies Physical Laws. Space is the medium of Space Resonances. For every type of wave phenomena we know in science, the properties of the wave media determine the behavior of the waves and their interactions with each other. SR waves in space follow this same rule. The laws of relativity, quantum mechanics, electron-charge, and energy transfer are derived from space properties, according to the three SR assumptions. There is also reason to suppose that baryons, the remaining particles, which are not fully understood in any theory, also derive from space. Therefore, it is a tentative conclusion that there is no basic entity except space itself, from which all observable properties of matter appear to be derived. Natural numerical constants have the same origin; electric charge e^2, velocity of light c, and Plank's constant h are properties of the fabric of space.

Speaking honestly, I find it hard to get used to these sweeping conclusions. Does everything really originate from only space? Do all the complex facets of our scientific world have only a single basis? Where has logic taken us? The idea is breath taking.

Understanding Relativistic Behavior. Space resonances provide an easy way to explain relativistic increase of mass (frequency). After grasping the fundamental fact that mass and frequency are identical, it becomes obvious that the Doppler frequency changes of the resonance IN/OUT waves are the same for two particles moving with respect to each other. Those frequency changes are the mass (frequency) increases of relativity. That is all there is to it.

Scoreboard for the Space Resonance. Let's make a list of the attributes of the SR theory.

It offers a logical mathematical explanation for the following laws, which have formerly been based solely on experimental observation:

The Conservation of Energy
Quantum Theory
Dirac Equation
Quantum Electrodynamics and Feynman diagrams
Special Relativity
Electric charge and Maxwell's Equations
Newton's Law, F = ma.

It provides plausible explanations for puzzles and enigmas of physics:

Dirac's two large number hypotheses
Mach's Principle
The origin of Gravity forces
The instant communication of the EPR experiments
The wave-particle duality of quantum mechanics
The mysterious charge cut-off of quantum electrodynamics
Conservation of Lepton and Baryon number

It offers a speculative cosmology (no wilder than others) to explain:

A non-violent origin of the universe
The mystery of the missing anti-matter.

Paradox Lost, Space regained. Explanations in science seldom remove all of the puzzling questions. Usually, several puzzling facts are replaced by another new puzzling fact which we have to swallow on the grounds that fewer puzzles are better than many. The space resonance theory removes many of the paradoxes of modern physics, but we are stuck with a new and strange idea of the IN-wave. How does it work? We don't know, but we can probably get used to it.

THE ONE REMAINING PARADOX:

Can we Prove the SR Theory Using the Outcome of Experiments?

The final test of any new theory is the successful outcome of an experiment whose results have been predicted by the theory. What experiments can be done to demonstrate the SR theory? Below are several possible experiments which may provide this evidence:

1. **Violation of Bell's Theorem may be a special key to the IN waves.** The apparent faster-than-light communication, found in the experiments to test Bell's Theorem, is the most striking enigma of quantum theory. The SR wave-model provides an explanation: The communication is carried by the IN waves of the space resonance. Repeating those experiments, in a somewhat different way, could demonstrate the existence of the IN waves.

This can be accomplished with an apparatus of the type used by Aspect, Dalibard, and Rogers (1982). Instead of making a random filter-setting during the passage of the photon to the other detector, the filter-setting should be made during the time just *preceding photon departure*. This is designed to frustrate communication by the IN waves. The result is predicted to be the linear correlation of *causality* as anticipated by Einstein in the EPR paper. Such an experiment would be direct evidence of the existence of the IN waves.

2. Lifetime of the Neutron. According to classical theory, it is generally believed that the lifetime of decaying particles or states is an inherent property of the particles themselves depending on their internal structure. Accordingly, regardless of how you measure a given lifetime, the same value is always expected. For example, the lifetimes of a neutron and atomic energy levels are fixed tabulated constants in the Handbook of Chemistry and Physics.

In contradiction, the space resonance concept predicts that the neutron (or other decays) may have different lifetimes under different conditions because it involves an energy exchange between the neutron and another external particle. Thus, the lifetime depends on the availability of external particles to receive the energy. In an ordinary Earth environment, the density of particles available for exchanges is fairly constant, so neutron lifetime measurements usually yield the same value. But if a lifetime measurement were made in a spacecraft located in "empty" space several lifetimes distant from the Sun, Earth, or other massive objects, the lack of readily available exchange partners would result in a longer lifetime. This should be true for any beta-decay lifetime. If this result were obtained, it would be direct evidence of the resonant energy exchange concept of space resonances.

3. Variation of Einstein's "A" Constant of Natural Emission. According to the classical ideas, an electron in an excited atomic state decays back to its normal state with a constant lifetime. The decay rate is given by Einstein's "A" coefficient. This rate is thought to be an inherent property of the atom and is expected to always be constant. The emitted photon is imagined to be free to travel forever and whether or not it becomes absorbed does not affect its emission rate.

According to the space resonance concept, the emission of a photon must be an exchange with another atom as a partner. Since the availability of partners can be altered by location or by appropriate reflectors of the radiation involved, the theory predicts that the "A" constant is not a constant but depends on the environment of the atom in the excited state. Although the average environment is very constant (which has led to the constant lifetime presumption) experiments can be done to alter this environment. For example, a decaying atom could be surrounded with a mirror which reflects the emitted radiation, thus altering its capability to exchange with a partner. The measured lifetime should then change. One could also alter the environment by isolating an excited atom in outer space, as was suggested for the neutron above. Any variation of the presumed "A" coefficient would be strong evidence supporting the transfer mechanism of SR energy.

Post Script (added by the editors): In May 1990, Dr. Wolff's third prediction, above, was verified when Dr. Herbert Walther experimentally discovered that the natural radiation rate of an atom was altered if surrounded by a resonant mirror cavity. Dr. Walther was awarded the Charles H. Townes medal at the CLEO Conference in Anaheim, California for his work.

The discovery was unexpected because scientists have always thought that the radiation rate of atoms and nuclei depended only on the "inner nature" of the particles.

In Dr. Walther's experiments, the atoms inside the cavity seem to know about the mirror's presence before its emitted photon gets there. This implies that the atom knows its own future. If the experiment had not been done, a suggestion that the atom knows its future would sound completely crazy.

4. Gravity of the Anti-particles. The SR theory suggests that anti-particles have anti-gravity. That is, a gravitational repulsion will occur between an electron and a positron, or between a proton and an anti-proton. This is difficult to measure, because the gravity force is so small compared to the electric force. However, it might be achieved if a few molecules of electrically neutral anti-matter were optically isolated. Or a spacecraft on the outskirts of our galaxy, equipped with particle detectors, could measure the streaming directions of anti-matter and matter. Are they streaming to or *from* our matter-composed galaxy? If anti-matter streams away from our galaxy, its gravitation is repulsive. This would favor the SR theory of the formation of the universe.

5. Measurement of Non-isotropic Inertia with Respect to our Galaxy. As mentioned in Chapter 11, the amount of matter in the universe is so immense that its effect on local inertia, according to Mach's Principle or to the *Space Density Assumption,* must be almost completely constant throughout our galaxy. But almost is not complete. The mass of our galaxy is distributed in an elliptical disk so a difference of inertia may exist for directions perpendicular and parallel to the plane of the galaxy. An estimated calculation suggests the variation could be 1 part in 10^7.

The instrumentation laboratory of MIT and several aerospace companies have developed inertial measuring instruments which can detect variations this small. If measurements of inertial acceleration should show *any* differences that correlate with the plane of the galaxy, this is strong evidence of Mach's Principle and the *Space Density Assumption* of the SR theory.

Predicted Theoretical Calculations Could Also Provide Evidence

The experimental predictions above are not the only proof of the SR theory. The theory also predicts the outcome of certain theoretical calculations. If they are tried and turn out as predicted, that is evidence, too. Some of these are as follows:

6. Does the SR Wave Equation Contain a Stable Nuclear Resonance? The existence of a stable nuclear resonance obtainable from the SR wave equation mentioned above was only an educated guess. Solving an equation in which the wave propagation properties depends on the amplitudes of the solutions is difficult, but not impossible. A procedure using successive iteration on a computer is always possible. If a solution is obtained, it can then be analyzed to see if it possesses the properties of a proton.

7. Do two SR Atoms Possess Gravitational Attraction? The rough calculation of the gravity of space resonances given above was not rigorous. To do it properly, the IN/OUT waves of a minimum of four space resonances (the electrons and protons in two H atoms) must be calculated to see if indeed a perturbation by the Hubble expansion leads to the correct value of gravity. Further, it would be especially interesting to see if antigravity exists for an anti-H-atom.

8. What is the Mechanism of a Neutrino? Not much has been said here about the neutrino, since it is not prominent in Nature's particle zoo, but like the photon, its properties have been well measured and must be accounted for. In addition to transferring energy (frequency exchange), like the proton, it also invokes spin exchange. How is this done? Does the SR idea contain a mechanism to do this?

Nature and the Universe have Simplicity and Beauty

If you believe the space resonance concept, you can appreciate and marvel at the simplicity and beauty of the universe. First it is hard to believe that Nature is really so simple, then, if you accept the logic, the consequences are breath taking.

Imagine that the microscopic world of the spaces resonances could be enlarged for us to see. That world is filled with shimmering wave centers, each surrounded by translucent spherical shells of standing waves that gradually fade into the distance.

Some of these micro pearls are arranged in exquisite crystal patterns of myriad variety — such as the cubes, pyramids, rhombs, dodecahedrons and so on, that form solid minerals. Others form the long chains and spirals of life molecules. Some are floating; others are rushing about in torrents and waterfalls as chemicals react and energetic changes take place. All are in a state of shimmering perpetual motion which begin at tiny wave centers and reach out seemingly forever.

The flow of entropy through the universe creates a multitude of streams, spurts, and explosions as the Minimum Amplitude Principle continually seeks to calm and smooth the irregularities of wave amplitudes in the space of the universe. Sudden disturbances create spreading circles of reaction causing the standing waves of nearby resonances to march inward or outward as waves constantly adjust and readjust themselves to the motions of their neighbors.

Reflecting on the origin of this gigantic geometric display, one realizes it all begins with a single entity, the space resonance. And the only rules to create them are the properties of space, of which they are made. What is space? This is the ultimate mystery. Can space really be that "nothing" we see transparently with human eyes, but which our astronomers measure and deduce to occupy an infinity from ourselves to the edges of the universe?

We cannot compute exactly the wave pattern of even one resonance, because each one is affected by the minute variations of a myriad of others, but the resonance and its rules of behavior are simple. It is easy to understand what they are and how they form the universe and the fundamental laws of science.

The rules of space are akin to the ancient chinese game of "Go" which has a few simple rules for placing black and white stones on a board. Anyone can learn it in a few minutes, but the game is so complex that no one will ever master it completely. Like the game of Go, it may not have been necessary to grasp the almost infinite consequences of Nature's rules to set them into motion. Is there a master game builder in the universe? Why did he begin ours? And what will the next one be like?

Where Can We Go From Here?

Adventures in the future might luckily be simple, in contrast to much of present day physics and astronomy. Simple experiments are possible because the basic fundamentals of science affect all levels of phenomena, not just the output of expensive super-energy accelerators. And, it is the simple basics which we need most to understand. For example, experiments at very low temperatures frequently produce mysterious results which cannot be explained using current ideas. Some polarization experiments in ordinary visible optics are not yet explained. The Bell Theorum and EPR have not yet been explained. These enigmatic results may be a consequence of our limited understanding of quantum theory and the nature of a particle. Their solution may be simple and possible; provided someone has the curiosity and perseverance to try.

Astronomical data painstakingly gathered by dedicated observers, are available to everyone. Enigmas of the quasars, galaxies, the red shift, radio stars, and other puzzles are awaiting solutions. How can these results be put together into a logical, understandable framework? This is the cosmology game and it is always fascinating. Do you want to be a modern philosopher?

Good luck. Have fun.

Appendices

Mathemathical Appendix
References
Index
Publications by Milo Wolff

Mathematical Appendix

This appendix is intended for use by mathematically trained persons. If you are not a mathematician, do not feel that you will be left behind because you do not understand it. If you have carefully read the preceding chapters you should already have a firm picture of all the important concepts and ideas without added mathematics. On the other hand, mathematical relationships are important to help pioneers on the road to further exploration, as well as to establish with certainty some of the ideas which are only stated in the text.

Most of the mathematical proofs below are separated subjects which stand alone. They have the same bold-face headings as the related material of the text.

The Particle Waves Assumption

The first assumption of this model is that the waves of the space resonances are solutions of a scalar wave equation whose waves propagate in space with velocity c. This is an assumption that *space* has this property of propagating particle waves. There is no proof that this is true, but it can be believed if the results are more in agreement with experiments than are other theories.

I. PARTICLE-WAVES ASSUMPTION:

Space can propagate scalar waves, not directly observable, according to

$$\nabla^2 \Phi - \frac{1}{c^2}\left(\frac{\partial^2 \Phi}{\partial t^2}\right) = 0$$

where Φ is a continuous scalar amplitude with values everywhere in space, and **c** is the propagation speed.

This equation is similar to many other oscillatory equations found in nature. The assumption provides only a propagation equation for the particle waves.

Two solutions of the equation, an inward-moving spherical "IN wave," and an outward-moving "OUT wave" are combined to form a standing wave whose properties are investigated in this paper. This combination is a "space-resonance."

An important result is that the scalar wave-equation allows the amplitude to be finite at the center. Mathematically, this is only possible for a scalar wave, not a vector wave.

Spherical Wave Solutions

The two spherical wave solutions which form the space resonances will be obtained now for the case of a stationary resonance.

The wave equation when written in spherical coordinates, becomes

$$\frac{\partial^2 \Phi}{\partial r^2} + \frac{2}{r}\left(\frac{\partial \Phi}{\partial r}\right) - \frac{1}{c^2}\left(\frac{\partial^2}{\partial t}\right) = 0$$

where Φ is wave amplitude and **r** is radial distance.

This equation has two solutions for the amplitude Φ; one of them is a converging (IN) spherical wave and the other is a diverging (OUT) wave,

$$\Phi^{IN} = \frac{1}{r}\Phi_0 e^{(i\omega t + i\kappa r)} \quad , \quad \Phi^{OUT} = \frac{1}{r}\Phi_0 e^{(i\omega t - i\kappa r)}$$

They can be combined so that the amplitudes at r = 0 are opposite. This combination removes the infinity at r = 0. The combined wave is the difference of the amplitudes of the IN and OUT waves of a resonance, which is written

$$\Phi = \Phi^{IN} - \Phi^{OUT} = Ae^{(i\omega_i t + i\kappa_i r)} - Ae^{(i\omega_o t - i\kappa_o r)}, \qquad i = IN, o = OUT$$

where $\omega = 2\pi mc^2/h = \kappa c$ is the mass-frequency of the space resonance. The complex amplitudes A include the range factor of 1/r and are alike because of the symmetry of the IN and OUT waves.

The intensity of combined IN and OUT waves is the envelope of $\Phi*\Phi$,

$$\Phi*\Phi = \left(Ae^{-(i\omega_i t + i\kappa_o r)} - Ae^{-(i\omega_i t - i\kappa_o r)}\right) \times \left(Ae^{(i\omega_i t + i\kappa_o r)} - Ae^{(i\omega_i t - i\kappa_o r)}\right)$$

After multiplying and reducing, the intensity becomes,

$$\Phi*\Phi = 2A^2 - 2A^2 \cos\left[(\kappa_i + \kappa_o)r - (\omega_i - \omega_o)t\right]$$

The distinction between frequencies and wave numbers of the two waves can be used to investigate different properties, but if there is no motion of the resonance, then $\omega_i = \omega_o = \omega$ and $\kappa_i = \kappa_o = \kappa$, and the intensity reduces to

$$\Phi*\Phi = 4A^2\left[\tfrac{1}{2} - \tfrac{1}{2}\cos(2\kappa r)\right] = \left[2A\sin(\kappa r)\right]$$

which is the envelope of the oscillating standing particle waves. The standing wave has nodes located at $r = n\pi/\kappa$. It has spherical symmetry in its own inertial frame, and may be envisioned like layers of spherical, concentric, oscillating nodes whose intensity decreases as $1/r^2$ away from its center.

The intensity at the center is obtained by taking the limit as $r \to 0$ in the sine function and in A. It is equal to the constant part of A, so the absurdity of

MATHEMATICAL APPENDIX

the point charge of the electron leading to an infinite energy does not occur. The standing waves in this spherical geometry are mathematically analogous to standing waves in a long pipe, except that the pipe has undergone a transformation that opens one end to 4π radians (becomes a sphere) while stretching the radius to ∞, and at the same time shrinking the other end of the pipe to a point at the origin.

Two Resonances With Relative Motion

Relative motion between two resonances is very important since both QM and special relativity are physical laws which depend upon the relative velocity. Accordingly we now investigate the properties of the resonance waves which arrive from another resonance having relative velocity $\beta = v/c$ with respect to the first.

The frequency of the resonances have been chosen to be equal to the mass-frequency of a fundamental particle like an electron. By doing this, the relation between a space resonance and a particle can be quickly seen. This is an assumption, or can be regarded as incorporating experimental measurement of masses into the theory.

The appearance of the waves from a resonance, which arrive at another resonance, are changed if relative motion with velocity $\beta = v/c$ exists. Then the Doppler effect alters the received frequencies, velocities, and wave numbers. The IN waves are red shifted and the OUT waves are blue shifted according to the relativistic Doppler factors, D and 1/D, where

$$D = \gamma(1-\beta), \quad 1/D = \gamma(1+\beta), \quad \text{and} \quad \gamma = (1-\beta^2)^{-1/2}$$

The effect is perfectly symmetrical as it must be in relativity. Both resonances receive the same information from the other because the relative velocity is the same for both.

Note that we are only calculating waves which pass by and through each resonance. There is no reception or interaction by either resonance, since we have not yet introduced any means of energy exchange.

Using these factors, the received Doppler-shifted wave amplitude is then

$$\Phi = A e^{i(ct+r)\kappa/D} - A e^{i(ct-r)\kappa D}$$

Inserting the expressions for the Doppler factors,

$$\Phi = A e^{i(ct+r)\kappa/\gamma(1+\beta)} - A e^{i(ct-r)\kappa\gamma(1-\beta)}$$

Multiplying exponents, rearranging, and factoring,

$$\Phi = Ae^{i\kappa\gamma(ct+\beta r)}\left(e^{i\kappa\gamma(\beta ct+r)} - e^{-i\kappa\gamma(\beta ct+r)}\right)$$

Combining the last two exponential terms, the Doppler-shifted wave of either resonance, received at the other resonance, becomes

$$\Phi = 2Ae^{i\kappa\gamma(ct+\beta r)}\sin\left[\kappa\gamma(\beta ct+r)\right]$$

This equation has the form of an exponential carrier wave modulated by a sinusoid. The surprising characteristics of the carrier wave are:

wavelength = $h/\gamma mv$ = deBroglie wavelength

frequency = $k\gamma c/2\pi = \gamma mc^2/h$ = mass-energy frequency

velocity = c/β = phase velocity.

The modulating sine function has:

wavelength = $h/\gamma mc$ = Compton wavelength

frequency = $\gamma mc^2\beta/h = \beta$ (mass frequency) = momentum frequency

velocity = βc = v = relative velocity of the two resonances.

These wave properties are complete. They show that the two resonances contain the law of special relativity, and the law of QM implied by the deBroglie wavelength. All the parameters that can be measured for a moving particle, are contained in the above equation. Respectively, they are: The two quantum-mechanical parameters: a deBroglie wavelength and the Compton wavelength; and for relativity, the elements of the four-momentum vector: i.e. rest mass and three components of linear momentum. The latter are expressed in terms of frequency with the correct Lorentz factors.

Because the resonances contain the deBroglie wavelength, it can be used to obtain the Schroedinger Equation as originally constructed by Schroedinger.

When Maxwell propounded his four equations in 1886, they were heralded as the fundamental expression of electromagnetic laws. Later, 1905 and later, Einstein's special relativity was discovered, but it was not significantly recognized for about a half-century that the two laws of induction

Magnetic Equations from Coulomb's Law and Relativity

involving magnetism were a consequence of relative motion between charges, or that relativity played a role.

Even today, few scientists are fully aware that magnetic forces are a perturbation of the Coulomb electric forces as a result of relative motion. Maxwell's Equations, despite their fundamental value in the study of electricity, are not fundamental laws of Nature. Only the Coulomb force law and the special relativity are fundamental. Maxwell's Equations are obtained from these two.

In order to make it clear that Maxwell's magnetic equations do not belong on the list of fundamental laws, I show below how they are obtained from Coulomb's law and relativity.

Transformations of Special Relativity

We will use transformations between a reference frame 1 and a reference frame 2 which have a relative velocity v in the x direction. Subscripts 1 and 2 on the symbols indicate that they represent quantities measured by observers located in frame 1 and frame 2.

The relativistic expansion factor γ is

$$\gamma = \frac{1}{\sqrt{1 - v^2/c^2}}$$

Charge and Current Density. Charge density ρ_2 which is stationary in frame 2 is seen as a current density J_{1x} moving in the x direction in frame 1, and vice-versa. The transformation is

$$\rho_1 = \gamma \left[\rho_2 + \left(\frac{v}{c^2} \right) J_{2x} \right] \qquad (A-1)$$

Partial Derivatives. The operations of taking derivatives in one frame transform to the other frame as

$$\frac{\partial}{\partial x_1} = \gamma \left[\frac{\partial}{\partial x_2} - \left(\frac{v}{c^2} \right) \frac{\partial}{\partial t} \right] \qquad (A-2)$$

$$\frac{\partial}{\partial y_1} = \frac{\partial}{\partial y_2} \quad , \quad \frac{\partial}{\partial z_1} = \frac{\partial}{\partial z_2}$$

Components of Electric Field E.

$$\begin{aligned} E_{x2} &= E_{x1} \\ E_{y2} &= \gamma \left(E_{y1} - v B_{z1} \right) \\ E_{z2} &= \gamma \left(E_{z1} - v B_{y1} \right) \end{aligned} \qquad (A-3)$$

Force Transformations ($v_2 = 0$)

$$F_{x1} = F_{x2} \qquad F_{y1} = \frac{F_{y2}}{\gamma} \qquad F_{z1} = \frac{F_{z2}}{\gamma} \qquad (A-4)$$

Lorentz Transformations:

$$x_2 = \gamma(x_1 - vt_1)$$
$$y_2 = y_1 \qquad \text{and} \qquad z_2 = z_1 \qquad (A-5)$$

How to Derive Maxwell's Curl B Equation

We want to show that the equation of electric induction

$$\nabla \times B = \frac{J}{\varepsilon_0 c^2} + \frac{1}{c^2}\left(\frac{\partial E}{\partial t}\right)$$

can be obtained from equations (A-1, A-2, and A-3) and Coulomb's law.

Choose that the density of charge ρ_2, is stationary in frame 2, moves with velocity v with respect to frame 1. Therefore in frame 2, the magnetic field $B = 0$.

From eqn. (A-1), we see that $\rho_2 = \rho_1/\gamma$.

Using Gauss's law (vector form of Coulomb's Law), $\nabla \cdot E_2 = r_2/e_0 = r_1/e_0\gamma$, we can write in cartesian coordinates,

$$\frac{\partial E_{x2}}{\partial x_2} + \frac{\partial E_{y2}}{\partial y_2} + \frac{\partial E_{z2}}{\partial z_2} = \frac{\rho_1}{\gamma \varepsilon_0}$$

Then from the transformation equations (A-2) for derivatives and from (A-3) for E components, we can find the field in frame 1,

$$\nabla E_1 + \frac{v}{c^2}\left(\frac{\partial E_{x1}}{\partial t}\right) - v\left(\frac{\partial B_{z1}}{\partial y} - \frac{\partial B_{y1}}{\partial z}\right) = \rho_1 \frac{(1-\beta^2)}{\varepsilon_0}$$

We can replace the first term using, $\nabla \cdot E_1 = \rho_1/\varepsilon_0$, which cancels the next to last term, and then divide by v to get,

$$\frac{1}{c^2}\left(\frac{\partial E_{x1}}{\partial t}\right) - \left(\frac{\partial B_{z1}}{\partial y} - \frac{\partial B_{y1}}{\partial z}\right) = \rho_1 \frac{v}{c^2 \varepsilon_0}$$

The term in brackets is the x component of curl $B_1 = \nabla \times B_1$. Corresponding terms for the y and z components can be obtained by rotating indices. Adding

MATHEMATICAL APPENDIX

all three components, replacing $\rho_1 v$ by its equivalent J_1, dropping the subscripts 1, and rearranging terms, we obtain

$$\nabla \times B = \frac{J}{\varepsilon_0 c^2} + \frac{1}{c^2}\left(\frac{\partial E}{\partial t}\right)$$

which is what we sought to show. QED.

How to Derive Maxwell's Curl E Equation

We want to show that Maxwell's rule for magnetic induction of electric fields by changing magnetic fields

$$\nabla \times E = \left(-\frac{\partial B}{\partial t}\right)$$

is a result of Coulomb's Law and the relativistic transformations (A-1, A-2, and A-3). That is, one need not use the notion of the magnetic field B or Maxwell's equation as the origin of the induced E field.

Choose two charges Q_c and Q_m moving in reference frame 2 at the same velocity v with respect to frame 1. Q_c is at the origin and Q_m is at the point $(x_2, y_2, 0)$. Viewed by an observer in frame 2, the charges are affected only by each other's Coulomb forces of attraction and repulsion; There are no magnetic fields present.

When viewed from frame 1 which can be our laboratory, one of the moving charges, say Q_c, is classically regarded as a current which produces a magnetic field B. The other charge Q_m will be affected by the magnetic field of moving Q_c according to the usual magnetic force rule,

$$F = Q_m(v \times B)$$

We want to get this result without using Maxwell's equation, by determining that the magnetic field B can be obtained from Coulombs law and the relativistic transformations.

Begin by writing the forces on Q_m due to Q_c using Coulomb's law as observed in frame 2,

$$F_2 = \frac{Q_m Q_c}{4\pi r_2^3}(x_2 + y_2)$$

where x_2 and y_2 are the vectors of position. Then, transform the components of this force to frame 1, using force transformations (A-4),

$$F_{x1} = F_{x2} = \frac{Q_c Q_m x_2}{4\pi\varepsilon_0 \left(x_2^2 + y_2^2\right)^{3/2}}$$

$$F_{y1} = F_{y2} = \frac{Q_c Q_m y_2}{\gamma 4\pi\varepsilon_0 \left(x_2^2 + y_2^2\right)^{3/2}}$$

and

$$F_{z1} = 0$$

We must also transform the lengths x_2 and y_2 into frame 1 at $t_1 = 0$ using the Lorentz transformations (A-5). Then, rearranging, and bringing $(\gamma)^{-2} = 1 - v^2/c^2$ into the numerator, one gets,

$$F_{y1} = \frac{\gamma Q_c Q_m y_1}{4\pi\varepsilon_0 \left(\gamma^2 x_1^2 + y_1^2\right)^{3/2}} \left(1 - \frac{v^2}{c^2}\right)$$

and

$$F_{x1} = \frac{\gamma Q_c Q_m x_1}{4\pi\varepsilon_0 \left(\gamma^2 x_1^2 + y_1^2\right)^{3/2}}$$

These two terms can be combined into a term with the force pointed along the radial vector $r_1 = x_1 + y_1$, between charges, plus a vector product term with the velocity, pointed in the z direction,

$$F_1 = Q_m \left[\frac{\gamma Q_c r_1}{4\pi\varepsilon_0 \left(\gamma^2 x_1^2 + y_1^2\right)^{3/2}}\right] + Q_m \left[v \frac{\gamma Q_c v y_1 k}{4\pi\varepsilon_0 c^2 \left(\gamma^2 x_1^2 + y_1^2\right)^{3/2}}\right]$$

The first term is the ordinary Coulomb force between the charges, and the second term may be recognized as the magnetic force = $Q\,(v \times B)$. The magnetic field B is the result of the apparent current $(Q_c)v$ arising from the relative velocity v of the observer and the charge Q_c. The constants in the denominator are identified as the inverse of the usual magnetic force constant $\mu_0 = 1/4\pi\varepsilon_0 c^2$.

We have obtained the magnetic force law from Coulomb's law and relativity which is what we sought to show.

MATHEMATICAL APPENDIX

PHYSICAL CONSTANTS – Measurements are the Backbone of Science.
Given in the table below are recent values of the more important measured physical constants. You may want to use these in your own calculations and explorations.

TABLE 13-1
PHYSICAL CONSTANTS (April, 1988)

Quantity	Symbol or Equation	Value
speed of light	c	299,792,460 m/sec
Plank constant	h	$6.626\ 075 \times 10^{-34}$ J-sec
elementary charge	e	$1.602\ 177\ 3 \times 10^{-19}$ Coul
electron mass	m_e	$9.109\ 389\ 7 \times 10^{-31}$ kg
proton mass	m_p	$1.672\ 623 \times 10^{-27}$ kg
		$= 1836.152\ m_e$
neutron mass	m_n	$1838.683\ m_e$
solar mass	m(sun)	1.989×10^{30} kg
gravitational constant	G	$6.672\ 59 \times 10^{-11}$ m^3/kg-sec^2
Avagadro number	N_A	$6.022\ 136 \times 10^{23}$ mol/mole
permittivity of free space	e_o	$8.854\ 187 \times 10^{-12}$ Coul^2/Nm^2
1/Hubble constant, H_o	$1/H_o = t_o$	1.5×10^{10} years
		$= 4.7 \times 10^{17}$ sec
Combined quantities		
fine structure constant	$a = e^2/2e_o hc$	1/137.0360
classical electron radius	$r_e = e^2/4\pi e_o m_e c^2$	$2.817\ 941 \times 10^{-15}$ m
electron Compton wave length	$h/m_e c = r_e/a$	2.42631×10^{-12} m
Bohr magneton	$m_B = eh/2m_e$	$5.788\ 382 \times 10^{-11}$ MeV/Tesla
critical density of universe	$3H_o^2/8\pi G$	1.9×10^{-26} kg/m^3

MATHEMATICAL APPENDIX

References

H. L. Armstrong, 1983, *Am J. Phys.* 51,103.

A. Aspect, J. Dalibard, and G. Rogers, 1982, *Phys. Rev Ltrs.* 49, 1804.

J. S. Bell, 1964, *Physics* 1, 195.

D. Bohm, and B. Hiley, 1984, *Found. of Phys.* 14, 255.

J. F. Clauser, and A. Shimony, 1978, *Rpts on Prog. in Phys.* 41, 1881.

M. Bunge, 1973, **Philosophy of Physics** p107-110 (Reidel Pub. Co.).

B. d'Espagnat,1983, **In Search of Reality** (Springer-Verlag)

P. A. M. Dirac, 1929, *Proc. Roy. Soc.* A 117, 610.

P. A. M. Dirac, 1937, *Nature*, London, 174, 321.

A. Einstein, B. Podolsky, and N. Rosen, 1935, *Phys. Rev.* 47, 777.

W. Heitler, 1944, **The Quantum Theory of Radiation** (Oxford U. Press) 1944.

N. Herbert, 1985, **Quantum Reality** (Doubleday, Garden City, NY)

E. Hubble,1936, **The Realm of the Nebulae** (Oxford U. Press).

H. Pagels, 1982, **The Cosmic Code** (Simon and Schuster)

E. Schroedinger,1931, *Sitzungsb. Pruess Akad. Wiss. Phys-math* 3, 1.

H. Tetrode, 1922, *Z. fur Physik* 10, 317-21.

J. A. Wheeler,and R. P. Feynman, 1945, *Rev. Mod. Phys.* 17, 157.

G. Zukav, 1979, **The Dancing Wu-li Masters** (Morrow, NY)

Index

A age of the universe 165
   angular momentum 58
   antimatter 170, 225
   Aristotle 9, 101
B baryons 152, 159, 170,
   Bell's theorem 140, 206, 210
   big-bang theory 174-6
C calculus 28
   charge, fundamental 146, 193-4, 216
   Clifford, William K. 78, 98
   cloud chamber 41
   closed universe 167
   complex numbers 75, 118
   conservation laws 168-70
   constants, physical 246
   Copenhagen Doctrine 12
   cosmology 162, 167, 222
   cosmos realm 20
   Coulomb's Law 154
D dachshund universe 223
   Dirac, Paul 131
   deBroglie, Louis Duc 6, 128, 184
   density of matter 168
   Dirac numbers 173, 202, 206
E Eddington, Arthur 78, 223
   Einstein, Albert 78, 127, 140

electromagnetism 153-56, 226
electron diffraction 128
electron paradoxes 147, 193, 214
energy conservation 51, 168, 191
energy-mass equivalence 52
energy transfer 190-2, 207-10
exclusion principle 148
expansion of the universe 164-5
F Feynman, Richard 188
   fine structure constant 246
   flatland 69
   forces, unity of 192
G galaxies 163-6
   Galileo, G. 9
   Gauss, Karl and geometry 99
   general relativity theory 102
   gravitation, Newton's Law of ... 25, 50
   gravity, anti- 171
   gravity, cause of, SR 202
   group theory 79, 96
H Heisenberg's uncertainty
       principle 138-9
   Hubble, Edwin 164
   Hubble's constant 165
   human realm 20, 126
   Hydrogen,
       density in universe 165

- Hydrogen wave functions 122, 214-16
- **I** inertia force 25, 53, 203
- **K** Kepler, Johann 46
- **L** laws, fundamental 32, 87, 156
  - leptons 152
  - lineland 68
  - Lorentz, H. 6
- **M** Mach, Ernst 171-2, 203
  - magnetic field 155
  - magnetic poles 73
  - mass, inertial 25, 202
  - mass, gravitational 25, 203-5
  - Maxwell's Equations 153-6, 243-5
  - Minimal Amplitude Principle 191
  - Momentum, conservation of 56
- **N** Newton, Isaac 10
  - Newton's Second law (see inertia force)
  - Newton's Second Law, cause of, SR 203
  - nuclear forces 151, 196-8, 220
  - neutron 151
- **P** pair annihilation 153, 170, 183-4
  - particle concept 133, 150
  - Penzias, A. and Wilson, R. 174
  - photo-electric effect 127
  - photon 26, 127, 144, 195
  - Planck, Max 126
  - Planck's constant 52, 246
  - predictions 232-35
  - probability 88, 192
- **Q** quantum theory 120, 185, 192
  - quantum electrodynamics, QED 132
  - quantum realm 20
  - quarks 159
- **R** red shift 165
  - rotation 74
  - relativity, special 61, 72, 185
- **S** space, meaning of 94, 230
  - space density assumption 188
  - space resonance 178-229
  - spectra, optical 40
  - spherical coordinates 122
  - spherical rotation and spin 76, 96, 214-16
  - spin 147, 214
  - SU(2) 79, 96
  - symmetry 71
- **T** Tetrode, Hans 199
  - time, meaning of 24, 188, 221
  - time reversal 188
  - trigonometry 75
- **U** uncertainty principle 138-9
  - universe, Chapter on 162
  - universe, origin of, SR 224
- **W** wave equations 129, 182
  - wave-particle duality 136-7, 225
  - waves 106
- **Z** zero sum 169-70

Publications by Milo Wolff

(Two papers* reprinted in) ***Milestones of Optics - Polarization*** published by SPIE - International Society for Optical Engineering, Bruce Billings, Ed., 1990.

(Book) ***Exploring the Physics of the Unknown Universe - An Adventurer's Guide***, M. Wolff, Lib. of Congress No. 88-51875, 1990.

Calculating Rayleigh scattering from particulate surface and Saturn's rings, Milo Wolff and Auduoin Dollfus, *Applied Optics* **29**, 1496-1502 (1990).

Photopolarietry of Asteroids, Chapter in *Asteroids II*, 594-616, U. of Arizona. Press, Binzel et al, Eds., 1989.

Theory and Application of the Negative Branch of Polarization for Airless Planetary Objects, Auduoin Dollfus & Milo Wolff,. *Proceedings, 12th Lunar/Planetary Conf.*, Houston, May 16, 1981.

Computing diffuse reflection from particulate planetary surface with a new function,. Milo Wolff,. *Applied Optics* **20**, 2493-2498 (1981).

***Theory and Application of the Polarization-Albedo Rules,**. Milo Wolff,. *Icarus* **44**, 780-792 (1980).

(A test of the Wolff polarization model): "**Experiments to Test Theoretical Models of the Polarization of Light**", Geake, Geake, and Zellner. *Monthly Notices of the Royal Astron. Soc.* **210**, 89-112 (1984).

***Photopolarimetry of Scattering Surfaces and their Interpretation by Computer Model II**, final report. NASW-3201, Washington, DC. Milo Wolff, ITA Inc. (1979).

***Polarization of Light from Rough Planetary Surface,**. Milo Wolff,. *Applied Optics* **14**, 1395-1405. (1975).

Report of the Joint Pakistan-American Science Review Team, Birnbaum, Wolff, Welt, Hashmi, and Afghan, Sponsored by US National Science Foundation, Publ. Pakistan Science Foundation (1974).

Group Diagrams in Go, Milo Wolff, Amer. Go Journal **9**, No 2, (1974).

Direct Measurements of the Earth Gravitational Potential Using a Satellite Pair, Milo Wolff, *J. Geophys. Research* **74**, 5295-5300, Oct .13 (1969).

Precision Limb Profiles for Navigation and Research, Milo Wolff, *J. of Spacecraft* **4**, 978-983, August 1967.

A Position Tracer for Navigation in a Manned Satellite, Milo Wolff, *J. of Spacecraft* **4**,394-401, Mar (1967).

Planning Science Education for Development, Milo Wolff, *Asia Foundation Quarterly*, pp. 1-5, Dec. (1967).

A New Attack on Height Measurement of the Nightglow by Ground Triangulation, Milo Wolff. *J. Geophys. Research* **71**, No. 11, 2743-2748 June (1966).

Limb Spectrophotometry, Milo Wolff, *Research Report R-554*, Mass. Institute of Technology, August (1966).

The Profile of an Exponential Atmosphere Viewed from Outerspace and Consequences for Space Navigation, Milo Wolff, *Research Report E-1634*, Massachusetts Institute of Technology, September (1964)

An Optical Earth-Horizon Profile based upon Solutions of Chandrasekhar's Equations, Milo Wolff, *Research Report E-1687,* Massachusetts Institute of Technology, October (1964).

Gamma-Ray absorption in C^{12}, Milo Wolff and W. E. Stephens, *Phys. Rev.* **112**, 890, (1968).

A Pulsed Mass Spectrometer with Time Dispersion, M. M. Wolff and W. E. Stephens, *Rev. Sci. Instr.* **24**, No. 8 , Aug (1963).